101 THINGS EVERY KID SHOULD KNOW ABOUT THE HUMAN BODY

101 THINGS EVERY KID SHOULD KNOW ABOUT THE HUMAN BODY

By Samantha Beres

Illustrations by Mary Bryson

LOWELL HOUSE JUVENILE

LOS ANGELES

NTC/Contemporary Publishing Group

To Danielle
—S.B.

ACKNOWLEDGMENTS
The author would like to thank Penny Brandt; Sung Chang; Allison Gorham;
Shelley Gorham; Clare Kosinski; Shelia R. Murphy, R.N.,B.S.N.,OCN; and
Inese Wheeler, Ph.D., for adding their expertise to this project.

*This title has been reviewed and endorsed by Ross Kosinski, Ph.D., Professor of Anatomy
and Dean of Student Services, Midwestern University, Glendale, Arizona; and Eleanor
H. Smith, teacher, DuVal High School, Lanham, Maryland, and Educational Consultant,
Hope Educational Consultants, Lanham, Maryland.*

Published by Lowell House
A division of NTC/Contemporary Publishing Group, Inc.
4255 West Touhy Avenue, Lincolnwood (Chicago), Illinois 60712 U.S.A.

Managing Director and Publisher: Jack Artenstein
Director of Publishing Services: Rena Copperman
Editorial Director: Brenda Pope-Ostrow
Project Editor: Joanna Siebert
Typesetters: Carolyn Wendt and Treesha R. Vaux

Lowell House books can be purchased at special discounts when ordered in bulk for premiums
and special sales. Contact Customer Service at the address above, or call 1-800-323-4900.

Printed and bound in the United States of America

DHD 10 9 8 7 6 5 4 3 2 1

Library of Congress Cataloging-in-Publication Data
Beres, Samantha.
    101 things every kid should know about the human body / by Samantha Beres;
illustrated by Mary Bryson.
        p. cm.
    Includes index.
    Summary: Presents a variety of facts in such areas of anatomy as major organs,
major systems, the senses, diseases, and more.
    ISBN 0-7373-0222-4 (pa.: alk. paper).--ISBN 0-7373-0329-8 (cl.: alk. paper)
    1. Body, Human Juvenile literature. 2. Human anatomy Juvenile literature.
3. Human physiology Juvenile literature. [1. Body, Human. 2. Human anatomy.
3. Human physiology.] I. Bryson, Mary, ill. II. Title. III. Title: One hundred one things
every kid should know about the human body.
QP37.B466 1999
612--dc21   99-30621

INTRODUCTION
Your Body Is a Machine

If you look at cars or bicycles, you will notice that these machines are made of parts that work together to help them function. Your body is your own personal machine—it is made of many parts that all work together. Each part has a job to do. Next time you do something, like your homework, think about how your hands, eyes, and **brain** all work together to help you read and write. Notice how the different parts of your body are shaped, and where they are located. You would not be able to function the way you do now if your feet were where your hands are, or if your knees were shaped like your nose!

This book will teach you about the anatomy, or design, of your body. It will also teach you about the physiology of your body, or how all its parts and systems work together. You will learn a lot about the inside of your body and what role your internal **organs** play in your daily life. With all that your body does for you, it's your job to treat it properly: Take care of it through good nutrition, cleanliness, rest, and exercise.

Although we all appear different on the outside—except for identical twins—the human body is something that we all have in common. We hope that by reading this book you will learn more about yourself, and appreciate the amazing design and operation of your personal machine: your body.

Key terms for you to learn, shown in **boldface** type, are explained in the Glossary at the back of this book.

1 Every human body is made up of trillions of cells.

A snowman is built of trillions of tiny snowflakes stuck together. The human body is also made of trillions of tiny pieces. But instead of snowflakes, we are made of cells. Cells are the building blocks of all living things. But the cells your skin is made of are different from the cells that make up your **blood**. Your bones are made up of yet another kind. Even though there are different types of cells, they are still all cells. Therefore, they must have some things in common.

What is it that makes a cell a cell? Every cell is like a city. Cities can be very different from one another, but they still have things in common. All cities have boundaries around them. Each cell has a **membrane** around it, which acts like a protective wall. The **nucleus,** found in most cells, is like the cell's City Hall—it manages how the cell operates.* Like the busy streets of a city, the **cytoplasm** is the place where all the work of the cell is done. This is the material between the nucleus and the cell membrane.

Red blood cells are one exception. They do not have a nucleus.

SHAPES AND SIZES OF CELLS

Most cells are so small that they can only be seen with a microscope. In animals, an average cell is 0.0004 inch across. Shapes of cells vary. Some look like columns, others are cubes. The cells in your cheeks are flat, and **nerve** cells look like long threads.

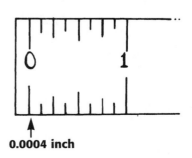

0.0004 inch

2 Cells have "engines" in them that generate heat.

Inside each cell, a structure that works like an engine makes energy and heat. Each "engine" is called a **mitochondrion.** Mitochondria are located in the cytoplasm of the cell. Each one helps you keep a steady body temperature.

The next time you get into a car, think about what happens to make the car's engine run. When the engine is running, fuel and oxygen cause a chemical reaction, which is a small explosion. Fuel keeps meeting with oxygen to keep the engine going. Your body works in a similar way, but instead of gasoline, your body uses **nutrients** for fuel. Blood picks up oxygen from your **lungs,** and nutrients from your digestive tract. Then the blood carries the nutrients and oxygen to your cells.

When a cell receives these supplies, it combines them to make a chemical reaction. This reaction makes heat. If you run around, you breathe harder and inhale more oxygen. Cells can add more oxygen to the fuel, work harder, and burn more to make more heat. This is why you get hot when you run around!

> # HANDS ON!
>
> **TAKE YOUR TEMPERATURE!**
> Ask an adult to help you take your temperature. Almost everyone has a steady 98.6°F body temperature. Your temperature could vary one degree above or below 98.6. Try taking your temperature at different times during the day—first thing in the morning, after running around, and again while sitting perfectly still. What are the results?

3 You started out as just one cell.

How did you come to be a whole person? It all starts when one of your father's **sperm,** which is like half a normal cell, meets one of your mother's **eggs,** which is also like half a cell. The two become a complete, whole cell, which later becomes an **embryo.** But how does that one cell

become a whole person? By a
process called **mitosis,** the
splitting of cells. Inside that first
cell are instructions, known as
DNA. These instructions copy
themselves, and then the cell
splits. Now there are two cells,
each with the same exact
instructions. The copying and
splitting continue. At first, all the

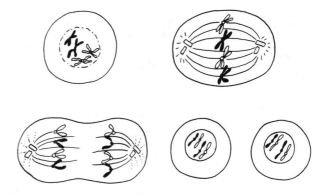

CELLS THAT HAVE SPLIT

cells are the same. Then cells of different shapes and sizes start to form.
Even though the cells change, each one still has its own set of instructions.
The copying of instructions and the splitting continue until there are
trillions of cells that form a complete **fetus** inside the mother's womb.

4 A group of cells makes a **tissue.**

Certain cells of the same type, such as blood cells or muscle cells,
form tissue. Blood cells and muscle cells would not join together to
make tissue, but a group of muscle cells would make muscle tissue.
   There are four basic types of tissue:
- Epithelial tissue mainly protects you. This type of tissue makes
  up your skin, and the lining of your mouth, throat, and
  digestive tract.
- Connective tissue holds things together and supports them.
  This type of tissue is found under your skin, in your **tendons,**
  and holds your organs in place.
- Muscle tissue is made of cells called muscle **fibers.** This type of
  tissue makes up your muscles and is attached to your bones.
- Nervous tissue consists of cells called **neurons**. Nervous
  tissue is found in your brain, spinal cord, and nerves—
  your nervous system.

**CLOSE-UP OF
MUSCLE CELLS**

BRIEF Bio

ROSALIND FRANKLIN (1920–1958):
Rosalind Franklin was a scientist who studied DNA. She realized that one way to get a better look at it was to crystallize it and then shoot X rays through the crystal. Her discovery helped scientists James Watson (1928–   ) and Francis Crick (1916–   ) identify the structure of the DNA molecule in 1953. They figured out that when a cell divides, the ladder of DNA pulls apart in the middle and unzips like a zipper. Unfortunately, Franklin was not fully acknowledged for her work until after her death.

CELLS JUST KEEP ON SPLITTING

Cells don't just stop splitting once a person is born. Right now, cells are splitting in your body! As some cells die, others divide to make new ones that replace the dead ones. Cells **regenerate** (split to make new ones) at different rates. Your skin cells and blood cells last only about 10–30 hours. They are constantly dividing to replace dead cells. The cells that make up some of your muscles live longer and will only divide to make more every few years.

WHAT IS DNA?

You may hear about police using DNA evidence to help solve a crime. DNA stands for deoxyribonucleic (dee-AHK-see-RYE-boh-new-KLEE-ik) **acid**. DNA is like a chemical "recipe" in every person's cells. The recipe determines the traits each person will have, such as the color of hair and eyes, height, body type—it's a very long list! The coded instructions of DNA contain enough information to fill a 300-page book! There are so many ways for the information to be organized that no two people in the world have the

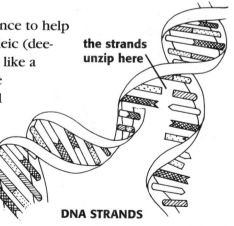

the strands unzip here

DNA STRANDS

same DNA, except for identical twins. This is why the police can be sure that if they have matched the DNA of hairs or skin left at a crime scene to the DNA of a suspect, the match is definitely right. The evidence can help prove the person was there.

5 A group of tissues makes an organ.

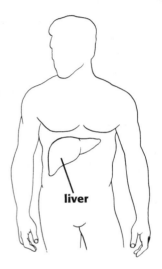
liver

Organs are made of a combination of two or more types of tissue. For instance, your **heart** is made up of muscle, connective, and nervous tissues. We often think of body parts, such as the kidneys and the **liver,** as organs. An organ is a group of tissues that do a certain job for your body. Your eyeballs are organs that allow you to see. Your legs are organs that allow you to walk. Your skin is an organ that protects you. Some would even say that just one bone is an organ.

Organs function in groups, or systems, to help your body run smoothly. For instance, your liver and kidneys work with your **stomach** and intestines to form your digestive system. Your brain, another organ, runs your nervous system. Sometimes one organ works with two systems. Your two kidneys are part of your excretory system and help your circulatory system by filtering blood. The **pancreas** is part of your endocrine system, and it releases **hormones** that allow the digestive system to work.

6 Some organs are needed for survival, others are not.

If there's a chance that someone's appendix is about to burst, a doctor will remove it. Doctors are not sure about the purpose of the appendix, so it is safer to remove one that might burst. In the past, tonsils were removed if a person was getting a lot of

FREAKY FACT!

A kidney transplant can cost anywhere from $25,000 to $130,000.

infections in the nose and throat. Now doctors recommend that tonsils be left in because they help fight **bacteria**. Your appendix and tonsils are organs you can live without.

The liver, heart, and pancreas, on the other hand, are organs that you need in order to live. You need at least one kidney and one lung in order to survive.

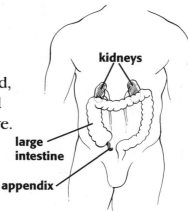

kidneys

large intestine

appendix

BRIEF Bio

JOSEPH E. MURRAY (1919–   ): Murray performed the first successful organ transplant in 1954, when he transferred a healthy kidney from one identical twin to another. He realized that the donors for organ transplants must have a genetic makeup similar to that of the recipient. Murray won the Nobel Prize for Medicine in 1990 with E. Donnall Thomas.

ORGAN TRANSPLANTS

As of January 1999, the patient waiting list for organ transplants included

TYPE OF TRANSPLANT	PATIENTS WAITING FOR TRANSPLANT*
Kidney	40,603
Liver	12,086
Pancreas	444
Islet of Langerhans	121
Kidney-pancreas	1,780
Intestine	199
Heart	14,175
Heart-lung	248
Lung	3,132

*Some patients are waiting for more than one organ.
Source: United Network for Organ Sharing (UNOS).

KIDNEY TRANSPLANTS

If a person has two failing kidneys, he or she may be able to have a new one put in. The procedure is called a transplant. It's not quite as simple as it sounds. A kidney must be donated from another person. In addition, the newly transplanted kidney will only work if the new kidney comes from a person with the same type of blood and tissue. Often, a family member will donate a kidney. If a recipient's body rejects the new kidney, the transplant will not be successful. Other organs, such as hearts and livers, are harder to come by. People with failing organs are on waiting lists that are years long.

7 Your body has several systems that work together like the parts of a machine.

A group of tissues and organs makes a **body system,** and many systems run a human body. Think of a bicycle and how all the parts work together. The pedals move the chain, which moves the wheels, which moves the bike. If you took away a wheel, or the chain, the bike wouldn't work!

Your body works like that, too. You can't remove one system from the body. They all work together. For instance, muscles can't be supported without the bones of the skeletal system. The circulatory system brings blood to the muscles and bones so they can stay healthy.

BODY SYSTEMS

SYSTEM	MAIN PARTS	FUNCTION/PURPOSE
Nervous system	Brain, spinal cord, nerves	Controls thought, movement, and other body systems
Endocrine system	Glands	Releases hormones
Respiratory system	Lungs, nose, throat	Carries air in and out of lungs
Circulatory system	Heart, veins, arteries, blood	Pumps blood around the body
Muscular system	Muscles, tendons, ligaments	Works with the skeleton to move
Skeletal system	Bones	Provides a framework for the body
Digestive system	Mouth, stomach, intestines, liver	Digests and absorbs food
Reproductive system	Testes, penis, ovaries, eggs	Makes new humans
Urinary system	Kidneys, bladder	Eliminates waste, maintains water and chemical balance
Sensory system	Eyes, ears, tongue, nose, and skin (hair and fingernails)	Provides protection

The nervous system and the endocrine system are the two most important systems. In a way, they control the other systems. For instance, the nervous system controls all the movements you make with your skeletal and muscular systems. And without a nervous system, you would not experience your sensory system, which includes your sense organs— your nose, tongue, skin, eyes, and ears. The endocrine system controls the hormones in your body. Hormones help the digestive system work and keep the blood in the circulatory system healthy—and much, much more.

BRIEF Bio

HIPPOCRATES (460–377 B.C.):
Hippocrates is known as the father of medicine. He grew up on the Greek island of Cos. The Greeks had a long medical tradition that helped educate Hippocrates. During his time, people were superstitious about disease, and thought healing depended on the gods. Hippocrates took a scientific approach to medicine. He looked for natural causes of illnesses and did not blame the gods. Without the help of modern techniques, he discovered how to set bones that were broken, as well as how to help a patient in recovery have a good attitude.

THE HIPPOCRATIC OATH

This oath, thought to have been written by Hippocrates, is a promise made by doctors. The promise is one of honesty in the practice of medicine, doing what is best for the patient, and keeping the patient's information confidential. Even thousands of years ago, doctors had a code of ethics. Doctors still take this oath today.

8 Your skin is the biggest organ of all.

Many people think of the skin as a body system, because it does such amazing things for the body. Do you know what your skin does for you?

■ Skin protects. It's like your own suit of armor. Your skin is waterproof. It keeps bacteria out of your body. When you sweat, your skin is helping you get rid of **waste**. Sweating is also one way for your body to cool itself off when it is hot.

■ Skin regulates temperature. When you're really hot, the blood vessels near the surface of your skin widen. This allows the warm blood to travel to the skin's surface to let heat escape. When you get cold, the vessels get narrow. This keeps the blood from getting too close to the skin and helps keep the heat in.

■ Skin provides touch. Touch is one of the five senses. Imagine a world without touch. You could put your hand on a hot stove and never know you are burning it! In this way, again, your skin is protecting you. Touch allows you to enjoy things, too, like the soft fur of a cat, or a silk shirt.

■ Skin takes in nourishment. When you go out in the sun, your skin absorbs sunlight. It is not safe to get too much sun, but you do need some. Skin turns the sun's ultraviolet rays into vitamin D. Vitamin D is needed to help the body absorb **calcium**.

GOOSE BUMPS

If you look closely at your skin the next time you have goose bumps, you'll see the hairs on your arm standing up. Our ancestors very likely had much more hair than we did. When they were cold and got goose bumps, their thick hair would stand up to catch more air, thereby keeping them warmer. Air, if it can be trapped, is a very good insulator. Our skin still reacts to the cold with goose bumps, but surely it doesn't keep us as warm as a sweater does! Many other **mammals** and birds get goose bumps, which fluff out their fur and feathers to trap air and help the animals stay warmer.

goose bump
epidermis {
dermis
hair follicle

9 Skin is made of two layers: the **dermis** and the **epidermis**.

If you look closely at your skin, you will see little openings called pores. You will also see hair. The pores and the root follicles of the hair start in the bottom layer of the skin, the dermis. In this layer are sweat **glands** and sebaceous glands. The sweat glands send sweat out through your pores, and the sebaceous glands produce an oil to protect and moisturize your skin. In the dermis, new skin cells are being made all the time. Your skin is alive! As the new skin cells are made, the old cells are pushed out to the surface where they die and eventually flake off. These old, dead cells are part of the epidermis, the top layer of skin.

FUN FACT! The thickness of your skin varies. It is thickest on the soles of your feet, and thinnest on your eyelids.

epidermis {
dermis {
sweat gland
fatty layer

pore of sweat gland
sebaceous gland
hair follicle
artery
vein
nerve

PROTRUDING PIMPLES

Pimples will start to appear as you approach your teen years. As you get older, the hormones will change in your body, and they will cause your skin to change. During this time, the sebaceous glands, which produce oil, get bigger and make more oil. These oils may clog your pores. If germs grow in clogged pores, you get pimples.

HANDS ON!

PRUNE A POTATO: Why do your fingertips get all wrinkled when you spend a lot of time in water? For the most part, your skin is waterproof. But if you stay in water long enough, your skin will eventually absorb some water. Parts of the skin swell up, causing your skin to look wrinkled, kind of like a prune. Watch how the skin of a potato does the same thing when it absorbs water. First, ask an adult to help you cut a cleaned potato in half. Put one half in water for at least an hour. Take it out and compare it to the dry half. What does the skin look like?

10 There are creatures called microorganisms living on your skin.

BRIEF Bio

ANTHONY VAN LEEUWENHOEK (1632–1723): Leeuwenhoek did many experiments with his own homemade microscopes. One day he scraped material from between his teeth and looked at it under a microscope. He saw what he called little "animalcules" moving around. When he reported his discovery, he didn't realize how important it was. At the time, many scientists did not believe him. Now we know that he discovered the bacteria that live on the human body.

Is your skin ever really clean? You may wash dirt and oils off in the shower, but billions of tiny, microscopic bacteria, yeasts, and fungi live on your skin. Your face, armpits, and neck have the most bacteria. Fungi thrive on your feet. Yeasts like to call the oily skin of your scalp home. What are all of these microorganisms doing on your skin? One thing they do is protect you from other organisms that cause disease.

WASHING YOUR HANDS

There's no way to get all the microorganisms off your skin, and it is natural for some of them to be there. It is still a good idea to shower regularly and wash your hands each time you use the bathroom and before you eat. Why? Because your hands come into contact with disease-causing organisms. If they are on your hands when you eat, the disease may pass from your food into you!

Keeping clean will help keep you healthy. If you are sick, you may put your fingers in your mouth, then touch something and spread the germs around. So it's a good idea to wash your hands a lot if you're sick or around someone who is sick.

11 No two fingerprints are alike.

Get a couple of your friends together and compare fingerprints. Not only will they all be different, but there is no one in the world with fingerprints

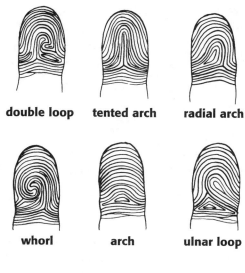

double loop   **tented arch**   **radial arch**

whorl   **arch**   **ulnar loop**

exactly like yours. Even the fingerprint patterns of identical twins are not exactly the same. Since everyone's prints are different, fingerprints are one way to identify individual people. This type of identification is used with accident victims, amnesiacs (people who have lost their memory and forgotten their name), and criminals.

There are some similarities among fingerprints, though. Fingerprints come in six general patterns. Which pattern is yours?

12 Pigments determine the color of your skin.

The color of a person's skin can fall anywhere in a range from very pale to very dark. What gives color to a person's skin—whether it be a brown, olive, or yellow tone—is pigment. Pigment is like a dye for the skin. A pigment called **melanin** is produced in

FUN FACT!

Carotene is another pigment that some people have in their skin cells. It adds a yellowish tone to the skin.

the cells in the skin's epidermis. Melanin provides color and absorbs the sun's ultraviolet rays.

If a person's skin is dark brown, his or her cells produce more melanin than the cells of a lighter brown person. And dark-skinned people can take in more sun without having to worry as much about getting a sunburn. The cells of pale-skinned people produce less melanin. **Albinos** have no melanin at all. The lack of melanin that albinos have is rare, and they must be extremely careful in the sun.

FUN FACT! Melanin also helps determine the color of your eyes.

INHERITED SKIN COLOR

Millions of years ago, our ancestors who lived in sunny places had darker skin with more melanin for protection. People who lived in colder places with less sun had lighter skin with less melanin. They didn't need as much protection from ultraviolet rays. Over the centuries, people moved to different parts of the world to live. You can look at more recent ancestors, like your parents and grandparents, to see where you inherited your skin color.

13 Hair protects your head and holds in heat.

Most of us have hair all over our bodies, except for the palms of our hands and the bottoms of our feet. And most people have a full head of hair. The hair on your head keeps you warm. Have you ever noticed how much warmer your whole body feels

FUN FACT! An average person has 100,000 hairs on his or her head.

HAIR FOR DOLLARS

In ancient Egypt, it was a regular practice for men to shave their heads bald. If they became sick and went into a hospital, they would stop shaving. At the end of a person's stay, he would shave the hair off and weigh it. The patient would be charged according to how much his hair weighed. If it was a long hospital visit, he would have grown more hair and would have to pay a higher fee!

if you wear a hat? Humans lose 5–30 percent of the body's heat through the head. This is because of the many blood vessels that are close enough to the scalp to allow heat to escape.

Your hair is made of cells. Each hair has a root in a **hair follicle** deep in your skin. In the follicle, new cells are growing all the time. The new cells of a piece of hair push the old cells out, causing the piece of hair to grow. As the cells reach the surface, they die. Therefore, the hair cells deep in the follicle are alive, but the hair you see, outside of your skin, is made of dead cells.

Like your skin, your hair is a certain color because of pigments, and this color is inherited from your parents. Each piece of hair has melanin. The more melanin the hair cells have, the darker the hair.

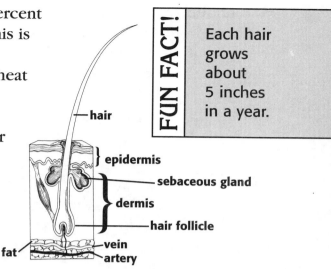

FUN FACT!

Each hair grows about 5 inches in a year.

hair
epidermis
sebaceous gland
dermis
hair follicle
vein
fat
artery

FREAKY FACT!

Baldness is hereditary and is linked to hormones. The hormones can cause the hair's root bulb to die. The root bulb is where new hair originates.

The kind of hair you have depends on the hair follicle. A round follicle makes straight hair, an oval follicle makes wavy hair, and a flat follicle makes curly hair. The follicle also has a lot to do with how fine or thick your hairs are.

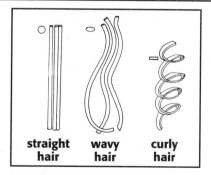

straight hair   wavy hair   curly hair

14 Fingernails and toenails protect your fingertips and toes.

What if you had to undo a knot without the help of your fingernails? When you stub your toe on something, it hurts. But if you didn't have toenails, it would hurt more. Fingernails and toenails protect your fingers and toes because of their hardness and because they have no feeling.

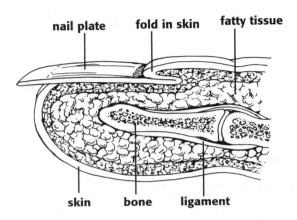

nail plate　　fold in skin　　fatty tissue

skin　　bone　　ligament

HANDS ON!

COMPARE FINGERNAILS AND HAIR: Hair and fingernails are made of the same material, keratin. But they sure do look different. Get a closer look. Take a small clipping of one of your fingernails. Also take one or two pieces of hair. Look at each under a magnifying glass. If you have access to a microscope, you'll be able to see lots more. What do you see? How are the hair and fingernail different or similar?

The skin underneath a nail is very sensitive. The nails grow out of a fold in the skin of fingers and toes. They are made of skin cells that are hardened by a **protein** called **keratin,** a material produced in the epidermis. As cells die, a layer of keratin forms. As layers add on, they push the older cells out and your nails grow. Keratin is also the substance that hardens the dead cells of your hair and the outside of your skin for protection. Nails are harder because they are made of thicker layers of keratin.

 ## Your muscular system consists of more than 600 muscles.

Everything you do takes muscle power. To walk, run, or jump, you need to use your muscles. But did you know that even when you're sleeping, your body is using muscles? When you breathe, muscles in your chest and **diaphragm** are working to allow your lungs to expand. Your heart is a muscle that works around the clock. And when you eat, muscles in your digestive system help the food move through.

In order to move, all muscles contract (get shorter), then relax (get longer). Bend your elbow and bring your fist toward you. The muscles on the inside of your arm contract. The outside muscles relax. Then, straighten your arm. The outside muscles contract and the inside ones relax.

BRIEF Bio

LEONARDO DA VINCI (1452–1519):
Leonardo da Vinci brought human anatomy and art together. By candlelight, the great artist would dissect bodies and sketch what he saw. He was the first person to dissect a human brain. And he even described how things, such as the brain and the muscles, worked. In order to illustrate human emotions, he would find subjects to draw who were fearful because they had been sentenced to death. He also drew pictures of people suffering from illnesses.

MUSCULAR SYSTEM

SORE MUSCLES

When you work your muscles, blood brings oxygen to them. Your muscles get sore if you work them harder than usual and they don't get enough oxygen. When they are worked and there isn't enough oxygen, they produce lactic acid. It is the acid in your muscles that makes them sore.

16 There are three types of muscles: smooth, skeletal, and cardiac.

Smooth muscles are hidden deep inside the body and quietly do their job to keep you alive. Smooth muscles are found in your digestive tract, for instance. They also line the walls of your arteries and **veins** to help pump the blood through. **Skeletal muscles** are different. They are attached to bones. There is only one **cardiac muscle,** which is the heart. It is a special kind of muscle and is part smooth and part skeletal.

Like everything else in your body, muscles are made of cells. The cells in

smooth muscle

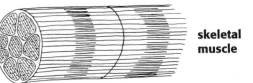
skeletal muscle

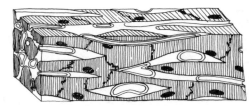
cardiac muscle

HANDS ON!

<u>LOOK AT SOME MUSCLE:</u> You will need to ask an adult if you can have a small piece of raw meat for this activity. Look at the meat under a magnifying glass or a microscope. What do you see? Can you see the fibers that make up the muscle tissue? If you can, compare different muscles. For example, if you are looking at chicken meat, observe a piece of muscle from the leg and compare it to the breast.

! SAFETY ALERT
Be sure to wash the meat, and wash your hands thoroughly before and after the above activity. Have an adult cut the meat for you.

muscles are fibers that bundle together to make up muscle tissue. But depending on what type of muscle it is, the fibers are arranged differently. Skeletal muscles, for instance, are cylinder-shaped fibers that are arranged in light and dark bands. Smooth muscles are not arranged in bands.

FUN FACT! Skeletal muscles account for 40 percent of your weight.

Muscles are not only made of different fiber structures, they come in different shapes. Some are flat, while others are long or short. The muscle around your mouth is shaped like a disc with a hole in the middle. Look at some parts of your body to figure out what the muscles inside must look like. Can you determine what shape they are?

17 Voluntary muscles make movements when you give them a command.

You go for a walk. You brush your teeth. You clean your room. These are all tasks that use voluntary muscles. Voluntary means that you think about the task, and then your muscles follow the command.

HANDS ON!

__WHAT DOES WHAT?__ There are a lot of spectacular things to notice about your muscles. If you pay close attention, you'll notice that moving your muscles in one place moves your body in another place. For instance, move your biceps and triceps, the two muscles in the upper part of your arm. If you flex them back and forth, the lower part of your arm bends and straightens. Now lift your arm. Even though it is your arm moving, the muscles moving it are in the shoulder area. Bend your knee to move the lower part of your leg. The muscles in the upper part of the leg do the work to move the lower part of the leg.

Most of the voluntary muscles are skeletal muscles. How do your muscles "hear" the command? It travels from your brain through nerves that are connected to single muscle fibers. The message causes some muscles to contract while relaxing other muscles, which makes a movement.

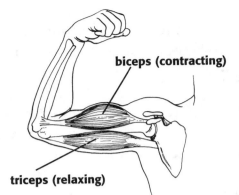

biceps (contracting)

triceps (relaxing)

18 Involuntary muscles move automatically.

When you run, you use voluntary muscles because you made the decision to run. But your heart, an involuntary muscle, starts working harder and your breathing automatically becomes deeper. Involuntary means that you don't command the motion; it happens automatically. In fact, the muscles that help you breathe work all day long without you even thinking about it.

Smooth muscles, which are part of your respiratory, circulatory, and digestive systems, are involuntary. For example, the digestive system is 30 feet long, and smooth muscles get the food to move from beginning to end. The **esophagus,** stomach, and intestines are not muscles, but they are lined with some involuntary muscles that contract and relax to move the food along.

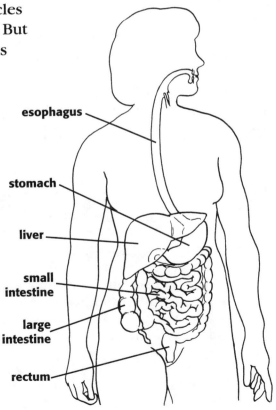

esophagus

stomach

liver

small intestine

large intestine

rectum

19 You breathe in and out with the help of the diaphragm muscle.

Your diaphragm is a sheet of thin, flat muscle that runs along the bottom of your lungs. It is arched until you inhale. When you inhale, your diaphragm contracts and flattens out, allowing your lungs to expand and take in air. Then it relaxes back into an arch and pushes up on the lungs, allowing you to exhale.

The diaphragm is the most important muscle for breathing, but it is not the only one. Your torso has two main parts, both of which contain muscles that help you breathe. The thorax, or chest, has many muscles that attach to the ribs and lift the ribs when you inhale. In the abdomen, abdominal muscles pull your ribs down, helping the lungs exhale. The diaphragm separates the thorax from the abdomen.

HICCUPS

Hiccups seem to come out of nowhere, but there is a reason for them. When the diaphragm is irritated, it contracts uncontrollably. This can happen if you eat too fast. The contraction, or twitching, causes the back of the throat to snap shut, which is what makes the sound of the hiccup.

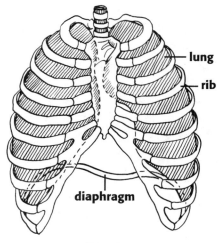

INHALATION

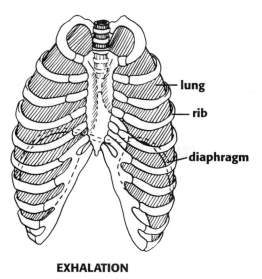

EXHALATION

20 Whether you're smiling, frowning, or chewing, it takes muscle.

Smile. Now frown. Imagine trying to make these facial expressions without moving the muscles in your face— impossible! The

FUN FACT!

Everyone shares some general expressions for emotions, such as happiness, surprise, anger, fear, and sadness. But don't be fooled. Even though all people share these expressions, there are 7,000 possible facial movements!

muscles in your head make facial expressions that are important for communicating without words. The muscles in your face work in pairs to make these expressions. Six pairs of muscles control just the area around your eyeballs, from blinking to squinting to smiling. There are also muscles on either side of your nose, which you can feel if you crinkle up your nose. Smile in the mirror, but watch the area around your eyes. You'll notice that it's not just the muscles around your mouth working. There are 17 muscles that help you smile! A frown takes 43 muscles.

The other really important muscles in your head are in your jaw and neck. The muscles in your jaw allow you to chew food. The human bite generates a force as great as 55 pounds. That means your teeth clamping down on a piece of food is the same as squashing it with a 55 pound weight. The molars in the back can make a force up to 200 pounds. The muscles in your neck allow you to look around and swivel your head in all sorts of directions.

EXPRESSIONS VERSUS GESTURES

Even though expressions are pretty much the same around the world, gestures are different. A gesture is a movement of the body that means something. For example, someone may put his hand up as a gesture to stop. You may point as a gesture to direct someone. The funny thing about gestures is that they differ from place to place. In Tibet, a country in Southeast Asia, a person will stick out her tongue if she's happy to see you. Not quite the same meaning as in the United States!

21 Tendons and **ligaments** are connective tissues.

Everything in your body is connected to something. Muscles are attached to bones, and bones are attached to **joints**. The tendons and ligaments help make these connections.

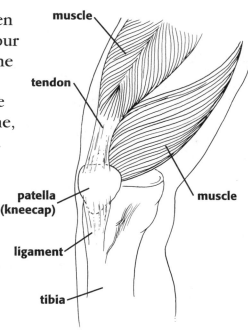

- Ligaments connect bone to bone. Often many ligaments will crisscross over your joints as reinforcement in attaching one bone to another. For instance, many ligaments in your knee hold your knee in place and attach your big thigh bone, the femur, to your shin bone, the tibia. Ligaments that are long, strong fibers can be found in places like your wrists, ankles, knees, and elbows.

- Tendons attach muscles to bones. Tendons allow the muscles to pull on the bones and spring back, which muscles need to do as the body moves. Tendons are like tough, elastic cords. They can be found all over your body.

THE ACHILLES TENDON

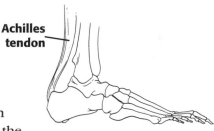

Your Achilles tendon is named after Achilles, a character from ancient Greek mythology. When he was a baby, his mother dipped him into the river Styx, which had the ability to make him powerful. Unfortunately, she held him by the heel, leaving it as a spot of weakness. When he grew up and became a warrior, he was shot in the heel with an arrow, which killed him. The Achilles tendon attaches your heel to your calf muscle. It is important because if it is cut or torn, you can't stand on your toes.

22 The skeletal system consists of 206 bones.

The skeleton of the human body is its framework. If you've ever seen a house before it is fully built, you saw a framework of strong beams underneath the walls and roof. You have strong bones that are your framework

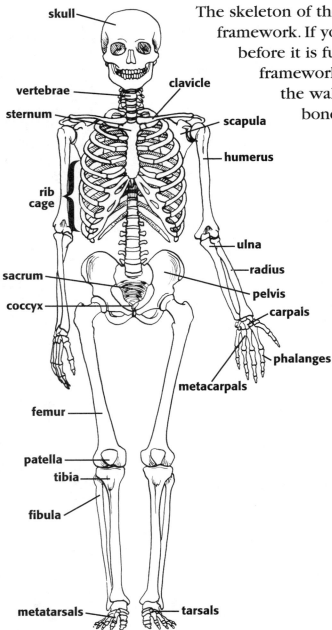

skull
vertebrae
sternum
clavicle
scapula
humerus
rib cage
ulna
radius
sacrum
pelvis
coccyx
carpals
phalanges
metacarpals
femur
patella
tibia
fibula
metatarsals
tarsals

BARE BONE NUMBERS

The younger you are, the more bones you have! While you are still growing, some of your bones will grow together and become one. Therefore, as an adult, you'll have fewer bones than you had as a child. Also, there are bones that are not counted as the 206 bones of an adult skeletal system. Sesamoid bones are small bones in the tendons that cover knees, hands, and feet. (Your knee bones, which are sesamoids, are one exception and are counted.) Wormian bones don't count, either. These are small bones in the joints in your skull.

and hold up your body. But bones do more than just hold you up. Bones grow and make you tall. They also work with the muscles to move you around. Some bones, such as the skull and rib cage, act as a protective shell for important organs such as the brain, heart, and lungs.

Humans and other mammals have a framework that is on the inside. This framework is called an endoskeleton. Some animals, like crabs and snails, have

FUN FACT! The smallest bones are in the ear and are less then one fourth of an inch long. The biggest bone is the femur, the thigh bone, which makes up one quarter of your height.

exoskeletons, which means that their framework is on the outside. A skeleton on the inside allows much more flexibility than an exoskeleton. Imagine trying to bend your arm if it had a shell on the outside of it!

23 | Joints allow you to bend and be flexible.

Bones are straight. Then how do we bend our elbows, knees, and other parts of our bodies? Joints, where the bones meet, can be found at places where parts of your body bend. Here are the different kinds of joints:

Hinge joints are in your elbows and knees. If you bend your elbow, you can see that your arm bends a lot, but only in one direction. Bend at your knee to see that the same is true of the bottom of your leg. It can only go back.

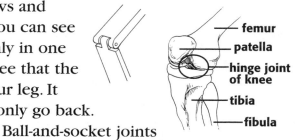

femur
patella
hinge joint of knee
tibia
fibula

Ball-and-socket joints are in your shoulders and hips. The ball-and-socket joints allow you to swing your legs while your hips stay perfectly still. These joints also allow you to swing your arms around while the shoulders stay still.

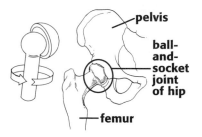

pelvis
ball-and-socket joint of hip
femur

Saddle joints are shaped like saddles. Saddle joints are in your wrists, thumbs, toes, and ankles.

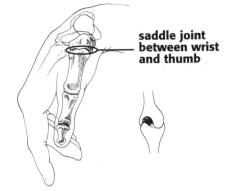

saddle joint between wrist and thumb

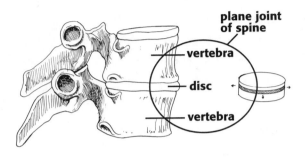

plane joint of spine

vertebra

disc

vertebra

Plane joints allow the bones to glide, but not with too much flexibility. Plane joints are in your **spine**.

Joints that don't move at all are called sutures. These joints hold the bones in your skull together.

FUN FACT!

What is that noise when you "crack" your knuckles or a joint? Sometimes air bubbles in the fluid between your bones pop. Other times, ligaments slide off a bone and make a noise like plucking a string.

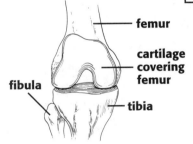

femur

cartilage covering femur

fibula

tibia

Joints work well because of the way the bones, tendons, and ligaments fit together. The ends of bones are also covered with a smooth **cartilage**. Some joints have a thicker piece of cartilage that acts like a shock absorber. A fluid keeps the joints oiled and also helps them move smoothly.

 # 24 Cartilage is an important part of the skeleton.

Human babies start out with a skeleton of 350 soft pieces of cartilage, instead of bones. That makes a skeleton with almost 150 more pieces than you will have as an adult! As newborns grow, the cartilage hardens into bone. Also, some of the bones grow together to make bigger, stronger bones. A child eventually has 206 hard bones, just like an adult. Some cartilage remains cartilage for a lifetime.

Cartilage is a rubbery substance that you can feel if you grab your ears or nose. By wiggling them back and forth, you can feel how flexible cartilage is. Aside from the ears and nose, cartilage is like a padding on the ends of your bones. It keeps the bones from rubbing together where they meet, such as at your knee and elbow joints.

CARTILAGE OF THE NOSE

ARTHRITIS

Osteoarthritis is common in elderly people but can occur in young people, too. Cartilage protects the bones from rubbing against one another. If this cartilage breaks down over time, the body cannot replace it, and arthritis develops. In advanced stages, the cartilage is so deteriorated that the bones start to rub each other. When the bones rub together, the outer membrane that covers the bone is irritated, and this can be extremely painful.

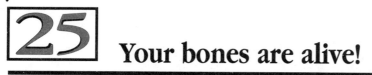

25 Your bones are alive!

We often see pictures that show smooth, curved shapes of white bones. They seem to be lifeless objects that have the sole purpose of holding up your body. But bones are not lifeless at all. They are made of living cells mixed in with **minerals,** which are nonliving. The minerals, one of which is calcium, are what give bones their hardness.

FREAKY FACT!

The most commonly broken bone is the clavicle (collarbone).

You may be most familiar with the hard, white outsides of a bone, which is called compact bone. Compact bone is covered with a thin, white membrane with blood vessels and nerves running through it. The inside of a bone, called cancellous bone, is quite different. It is still made of a hard substance, but with many nooks and crannies. The inside of a bone is like a sponge, making the bone strong but light. Nerves and blood vessels run through the insides of bones.

Bone cells are constantly going through a cycle of dying and regenerating (making new ones to replace the dead ones). Bones are made up of living cells. **Osteoblasts** produce new bone, and **osteoclasts** break down old bone. When you are young and growing, the osteoblasts make much more new bone than the osteoclasts dissolve. When you reach about the age of 20 and stop growing, the osteoblasts and osteoclasts continue their jobs at an evened-out rate. As the bone cells are dissolved, they are replaced throughout an adult's life. Most cells of bone are called **osteocytes** (bone cells), which are mature.

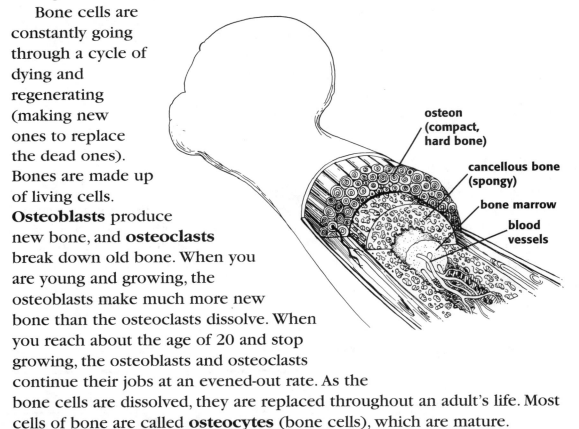

osteon
(compact,
hard bone)

cancellous bone
(spongy)

bone marrow

blood
vessels

HEALING BROKEN BONES

When a bone is broken, a doctor may have to set it so it will heal straight. Osteoblasts, new bone cells, form at the break, making a callus. The callus is like a patch that eventually becomes hard with minerals. A healthy, healed broken bone is as hard as it ever was.

26 When you are born, your skull is made of 28 bones.

Like a built-in hard hat, the skull protects one of your most important organs: the brain. But your hard hat is not fully developed when you are first born. In fact, the 28 skull bones of a newborn do not even touch one another. On the head of a newborn, there are soft spots between the bones. As the skull develops over the first 18 months, bones harden and some grow together. As an adult, you have fewer bones in your skull.

NEWBORN SKULL

A fully developed skull feels like one rounded bone, but it isn't. The flat, irregular bones are connected by suture joints. The top part of your skull is called the cranium. The cranium is made of eight curved bones and covers your brain. The front part of the skull is made of 14 bones that shape your face. The mandible, your lower jaw bone, is the only one of these 14 bones that can move.

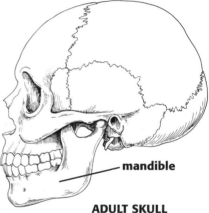

mandible

ADULT SKULL

HANDS ON!

FEEL YOUR RIBS: Your rib cage acts as a shield for your lungs and heart. You can feel your ribs to know how they are shaped. The 12 pairs of ribs are attached to the spine, and all but a couple curve all the way around to the front. In the front, 7 pairs are attached to a flat bone called the sternum. With your hands, follow from the sternum down. The ribs curve down and are held together by muscle and cartilage. The lower ribs that wrap only partway around you are only attached to the spine, or vertebral column. This whole structure protects your heart and lungs.

27 Your **spine** is made of 33 separate bones linked together.

If you run your fingers over your own backbone, you will feel many bumps. Each of these bumps is a vertebra, one of the bones that make up your spine. The 33 vertebrae are stacked on top of each other and held together with ligaments. Between each vertebra, cartilage acts to cushion the bones.

FUN FACT! The coccygeal bones are your tailbone. In humans, these bones run smoothly down the back. But in other animals, these bones are the beginning of a tail!

Your spine does help hold you up, but more importantly, it serves as a tube, called the spinal canal, for your **spinal cord**. The spinal cord runs from your brain through the canal.

Your backbone, or spine, works together as one unit. Different parts have different purposes:

- The top 7 vertebrae are the cervical bones that hold your neck and head up.
- Below these are the 12 thoracic vertebrae, which are attached to your ribs.
- The lumbar vertebrae in your lower back are the next 5.
- The 5 sacral and 4 coccygeal vertebrae come next. Even though your backbone is flexible, the sacral and coccygeal bones cannot move at all. These vertebrae fuse, or grow together, to make one solid bone by the time you're an adult. Then you'll only have 26 bones that make up your spine.

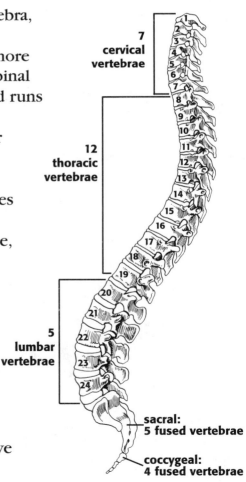

7 cervical vertebrae

12 thoracic vertebrae

5 lumbar vertebrae

sacral: 5 fused vertebrae

coccygeal: 4 fused vertebrae

HANDS ON!

MEASURE YOUR CHANGING HEIGHT: You are taller in the morning. It sounds like a myth, but it's true. Test it out. Measure yourself in the morning, right after you wake up. Stand flat against a wall and have someone mark your height with a piece of masking tape. Write "morning" on the tape. Then, right before you go to bed, take another measurement. Is the mark in the same place? Probably not. When you sleep, you are horizontal, so gravity is not pulling down on your spine vertically. But after you are awake and upright for hours, gravity pulls down on the spine, causing it to squish down.

VERTEBRATES VERSUS INVERTEBRATES

The animal kingdom can be divided into two groups: **vertebrates** and **invertebrates**. Vertebrates are animals with a backbone. Invertebrates are animals without a backbone. Humans and other mammals, plus reptiles and fish, are vertebrates. Invertebrates include animals like jellyfish and insects.

28 Bones can be classified into four categories: short, long, flat, and irregular.

Just by looking at yourself in the mirror, you can probably figure out that your leg and arm bones are long. Fingers and toes are also made of long bones. Short bones, shaped like cubes, are in your ankles and wrists. Ribs, shoulder blades, and skull bones are

FUN FACT!

All bones are connected to other bones, except for one—the hyoid bone at the top of the throat. It supports your tongue.

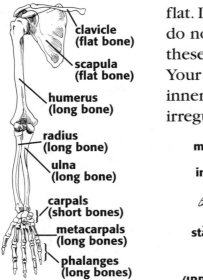

clavicle
(flat bone)

scapula
(flat bone)

humerus
(long bone)

radius
(long bone)

ulna
(long bone)

carpals
(short bones)

metacarpals
(long bones)

phalanges
(long bones)

flat. Irregular bones do not fit into any of these categories. Your vertebrae and inner ear bones are irregular.

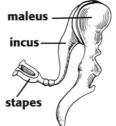

maleus

incus

stapes

**INNER EAR
(IRREGULAR) BONES**

FUNNY BONE

The feeling of hitting what we call a "funny bone" happens when you hit the end of a bone in your arm, the humerus. It feels like a knob at the side of your elbow. Hitting the end of the bone pushes on a nerve, which is what makes it feel funny—almost numb. But there is no real funny bone!

HIPS AND SHOULDERS—COLLECTIONS OF BONES AND JOINTS

If you grab one of your shoulders, you can feel the bumps and ridges under the skin. It's hard to tell what is going on there, but it is pretty complicated. A triangular-shaped bone called the scapula is your shoulder blade. The clavicle is the collarbone that goes from the front of your neck out. These two bones meet with the humerus, the upper arm bone, in a joint to form your shoulder.

The hip area is another complicated collection. Each side of the pelvis has a large curved, irregular bone called the ilium. The ilium, ischium, and pubic bone all join together at a joint where the leg meets the pelvis. The pelvis also joins with the sacrum, a bone at the bottom of the backbone.

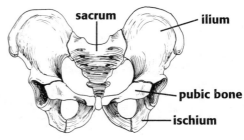

sacrum

ilium

pubic bone

ischium

There are 27 bones in each hand and 33 joints in each foot.

Where would you be without your hands and feet? Hands help you do delicate things, like pick an eyelash out of your eye. They also help you do everyday things, like write or make breakfast. Hands lend strength to lift and open things. You can even talk with your hands, if you know sign language. Hands are amazing! Each hand has 27 bones.

Just like your hands, your feet are a complicated jumble of bones, joints, and tendons. Even though feet do very different things than hands, they are structured in the same way. Some of the bones even have the same names. For instance, your finger bones and your toe bones are called phalanges. The bones that attach your fingers to your wrist are called metacarpals, and the bones that attach your toes to your ankle are called metatarsals. The foot also has cordlike tendons that run down the top of the foot. They work to lift the toes in the same way that tendons on the hand lift the fingers.

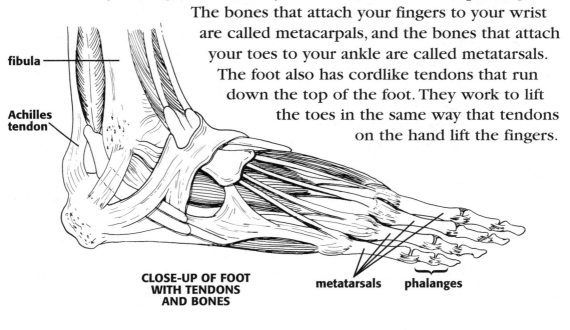

fibula

Achilles
tendon

**CLOSE-UP OF FOOT
WITH TENDONS
AND BONES**

metatarsals   phalanges

30 Humans are lucky to have an opposable thumb.

The Latin word for thumb is *pollex,* which means strong. The thumb allows you to grab on to things with strength. If the thumb was not able to press against, or oppose, the four fingers, we would not have the advantage of being able to pinch things, pick up things, and grab things! Not all animals share the convenience of an opposable thumb. Can you think of animals that have one?

HANDS ON!

APPRECIATE YOUR THUMB: This activity is fun to do with a friend. All you need is masking tape. First, have your friend tape each of your thumbs to the inside of your palm. Now, try to do a simple task, like tying your shoe or opening a door. (Do not do anything in which you could hurt yourself, such as cooking.) After you try one or two tasks, have your friend undo the tape. If your friend wants, he or she can try it next.

31 | The main parts of the respiratory system are the lungs, **larynx**, and **trachea**.

The role of the respiratory system is to exchange gases. What gases? Oxygen and carbon dioxide. We constantly inhale and exhale without even thinking about it. But what we are doing is taking in oxygen and exhaling carbon dioxide. When you inhale, air is drawn through the larynx and into

CPR

If someone stops breathing because of an accident, choking, or a stopped heartbeat, cardiopulmonary resuscitation (CPR) can save his or her life. Mouth-to-mouth resuscitation is the most widely practiced way to get a victim to breathe again. The victim is stretched out on his or her back, the neck is lifted, and a certified person will breath into the victim's mouth at regular intervals. If that doesn't work, a second person can start pushing down on the chest at regular intervals while mouth-to-mouth is being performed. Most people who practice artificial respiration have been trained and certified.

CPR

the trachea, the windpipe that leads to the lungs. Blood traveling through the lungs makes the gas exchange. As you inhale, your blood cells release carbon dioxide in your lungs to be exhaled. At the same time, they pick up oxygen in your lungs as you inhale.

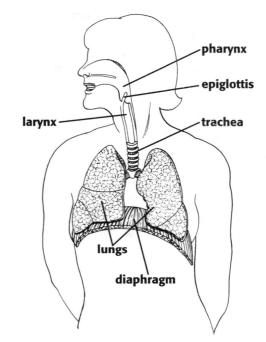

YAWNING

Yawning is often something we cannot control. But what does a yawn mean? It doesn't always mean that you're bored or tired. A lot of times, it means that you need more oxygen. A yawn causes your body to draw oxygen into your lungs.

32 Lungs are cone-shaped organs filled with tiny blood vessels.

You may picture your lungs as two hollow sacs that collect air. But lungs are not hollow at all! The lungs look like they are filled with layers of spongy material. Each lung

FREAKY FACT!

Around 30 percent of the adult population smokes. This group gets smaller each year, but the percentage of teens that smoke has risen in the past few years.

HYPERVENTILATION

Hyperventilating is taking in too much air. Your body needs only a certain amount of oxygen. But sometimes when a person gets nervous, or scared, he or she will react by breathing too heavily and taking in too much oxygen. This can cause a person to feel dizzy. The best thing for a hyperventilating person to do is breathe into a paper bag, to cut down on the amount of oxygen taken in.

has a **bronchus,** a large tube branching off the trachea. The bronchus feeds into **bronchioles,** which are many smaller tubes. These tubes keep branching off into smaller tubes that turn into **alveoli**.

Alveoli are sacs that are like little bubbles. These sacs are right next to **capillaries,** thin vessels that carry blood. The walls of the sacs and capillaries are so thin that oxygen can travel from the sac to the capillary and get into the blood. At the same time, carbon dioxide travels from capillary to sac and leaves as you exhale. So the air you breathe in makes it through the big tube, the trachea, passing through smaller and smaller ones until it can reach the blood.

SNORING

Snoring results when the soft palate in the back of the throat vibrates while breathing deeply during sleep.

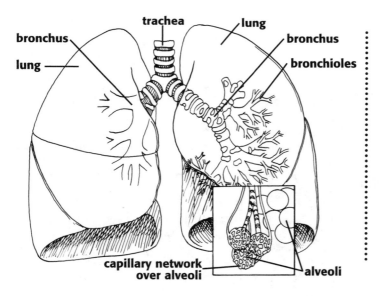

trachea
lung
bronchus
lung
bronchus
bronchioles
capillary network over alveoli
alveoli

THE AIR WE BREATHE

The air we breathe is not pure oxygen. In fact, too much pure oxygen could be poisonous. The air in our environment is about one-fifth oxygen. Sometimes in cases of emergency, doctors will give their patients pure oxygen.

THE DANGERS OF SMOKING

Long ago, people didn't know how unhealthy smoking was. Now the public knows that smoking is addictive, and that it can kill you. When a person smokes, he or she is inhaling poisonous gas called carbon monoxide. When this gas is inhaled, it goes into the bloodstream, like oxygen. Also, tar from cigarettes collects on the tissues in the lung when a person smokes. This irritates the lungs and can lead to cancer.

33 The circulatory system consists of your heart, veins, **arteries**, and capillaries.

The circulatory system transports blood and **plasma,** a liquid that makes up more than half of your blood, through the body in a "circular" path. You can imagine this system as a highway with many small roads coming off of it. Blood is the traveler on the highway of arteries, capillaries, and veins. Once blood makes the gas exchange in the lungs, the heart pumps it out through arteries. Blood travels all around your body making

Bio

WILLIAM HARVEY (1578–1657):
William Harvey, an English doctor, discovered how the heart, lungs, arteries, and veins work. Before his discovery, people believed that the blood moved through the body like the ocean tides—out and in—most likely through the same vessels. People also thought that air went through the blood vessels. Harvey studied hearts and lungs, arteries and veins, to find out how the circulatory system works. When he published his book *The Circulation of the Blood* in 1628, many people did not believe what he wrote.

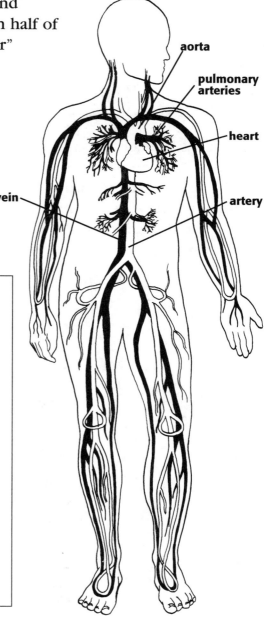

aorta

pulmonary
arteries

heart

artery

vein

oxygen dropoffs and carbon dioxide pickups. When it is low on oxygen, it hops into veins to return back to the lungs to drop off the carbon dioxide. There it starts all over again.

The movement of the blood is caused by your heart's pumping motion. As the heart contracts, it squeezes the blood out. As the heart relaxes, blood rushes in. The veins and arteries help with blood flow.

 The heart is your body's pump.

When the human fetus is just a cluster of cells at the beginning of its growth, the heart is one of the first organs to form. Once the body around the heart is complete, the heart gives life by pumping blood all the way down to the fingertips and toes!

FUN FACT!
The heart pumps 5–6 quarts of blood a minute through the circulatory system.

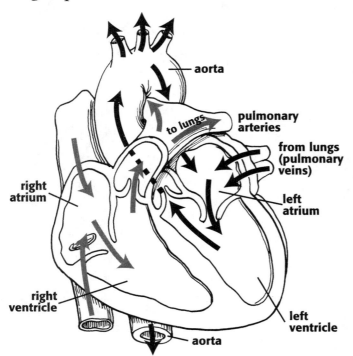

aorta

pulmonary arteries

to lungs

from lungs (pulmonary veins)

right atrium

left atrium

right ventricle

left ventricle

aorta

PACEMAKERS

If a person has a heart that stops or beats irregularly, a pacemaker will correct the heartbeat. A pacemaker can weigh 1–4.5 ounces and runs on a battery. Some pacemakers are implanted in the chest, and some are carried on the outside of the body. The pacemaker has an electrode that is threaded through a vein. When the heart fails to beat, the pacemaker sends an electrical signal to the heart to make the correction.

Your heart is constantly working. In one lifetime, a heart beats 4 billion times and pumps 600,000 tons of blood. If you put your hand to your chest, you will feel your heart beating. Now take your hand and make a fist. This is about the size of your heart. Contrary to our artistic idea of the heart's shape, it is really shaped more like a pear.

The heart actually has two pumps with chambers, or atria, in each one. As blood comes into the right **atrium,** it needs oxygen. It goes from the right ventricle to the lungs to get oxygen. Then it goes back to the left atrium, into the left ventricle, and then out the aorta full of oxygen.

HANDS ON!

TAKE YOUR PULSE: The heart beats an average of 70 times per minute. As the blood pumps through your arteries, it is pushed by the contraction of your heart. Each contraction is a beat. You can find your **pulse** at certain locations. Try placing two fingers on your wrist, or on the side of your neck. You can also try the back of your knee or the top of your foot. Do not use your thumb to take your pulse, because there is also a pulse in your thumb. Time the beats for 15 seconds, then multiply the number by 4. You should have a pulse rate somewhere between 70 and 100.

35 Heart disease is one of the leading causes of death in the United States.

Heart attacks are a result of heart disease. Heart disease comes in many forms. **Coronary artery** disease is the most common. When one or both coronary arteries become blocked, coronary artery disease results. The coronary arteries are the two blood vessels that bring fresh blood to the heart. If one of the arteries is blocked, part of the heart could die because not enough blood can get through, resulting in a lack of oxygen to the heart.

A coronary artery gets clogged when fatty deposits attach themselves to the vessel's lining. Once the artery is clogged, it becomes inflexible and narrow. This can cause more problems with blood flow, even leading to the formation of blood

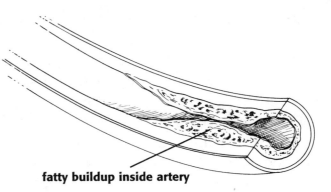

fatty buildup inside artery

clots, which could clog the artery completely. If this happens, the outcome is a heart attack. Eating fatty foods and smoking can cause heart disease.

HIGH CHOLESTEROL

Scientists have linked heart disease to very high levels of cholesterol in the blood. Cholesterol is a fatty compound found in human blood and tissue. It is made in the liver and is needed to help make new cells. We eat cholesterol in meat, butter, milk, cheese, and eggs. Having a high amount of cholesterol in the blood may cause fatty deposits in the arteries, leading to heart disease.

Everyone's body processes these substances differently. Therefore, someone might eat a diet high in cholesterol, but may not necessarily have high cholesterol levels in the blood. Also, someone may be very careful and eat a low-cholesterol diet, and may still have high blood cholesterol levels.

36 | Veins, arteries, and capillaries are "roads" for your blood.

If you think about driving to a city or a town, you may take a highway to get there. Then, to get to a certain place, like someone's house or a store, you take a regular road, or even a small side road. In your body, veins and arteries are like the highways and capillaries are like small side roads. Blood, the traveler, goes through the arteries to get to the main parts of your body, like your legs and arms. It exits onto the capillaries to get to specific places, like your muscles, your skin, and your bones. Veins are the highway home, back to the heart.

THE MEANING OF STROKE

If a blood clot or air bubble gets stuck in an artery and blocks the supply of blood to a portion of the brain, a stroke occurs. The lack of oxygen to a particular part of the brain will cause that part to stop working. Also, the functions of the body formerly controlled by the now-damaged part of the brain will be impaired. Some symptoms can be paralysis (inability to move), loss of speech, and unconsciousness. Strokes can even cause death.

Veins and arteries are not the same. Arteries always carry the blood away from your heart. They are thicker and more elastic, so they can stretch as pulses of blood pumped from the heart throb through them. The veins return blood to the heart. The veins are not as thick and stretchy as arteries, and they have valves, which are like trapdoors that only open one way. The valves make sure that the blood only goes in one direction, toward the heart. These "trapdoors" come in handy when the blood is traveling upward, against gravity.

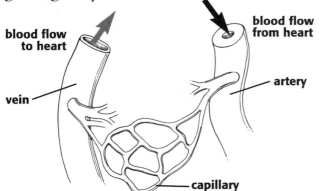

blood flow to heart

blood flow from heart

vein

artery

capillary

VEINS

At any given time, veins are holding more than twice as much blood as arteries. In fact, veins hold 70 percent of the blood. The pressure of the blood traveling through the veins is lower, causing the blood to move more slowly than through the arteries. Also, because veins have thinner walls, they can stretch out to accommodate the blood.

HANDS ON!

LOOK AT YOUR OWN BLOOD VESSELS: Look in a mirror in a well-lit room. Open your mouth and lift up your tongue. You should be able to see all sorts of blood vessels. The thick blue lines are veins, the thick pink lines are arteries, and the tiny little lines are capillaries. You can also see capillaries if you pull down your bottom eyelid.

37 There are four basic types of blood.

Human blood is classified into four general types: A, B, AB, and O. Why is it important to know what kind of blood a person has? Blood type must be known in case a blood transfusion is ever needed. Blood transfusions are given to accident victims who have lost a lot of blood. In the 17th century when doctors first started giving people blood transfusions, many people died. At the time, doctors did not know about different blood types. They also didn't know that certain blood types are not compatible. For instance, a person with type B blood cannot get a transfusion from a person with type A blood. The body would reject it, and the person would die.

FUN FACT! The first blood bank in the United States opened in 1940 in New York City.

BLOOD TYPES

| O+ | O- | A+ | A- | B+ | B- | AB+ | AB- |

BRIEF Bio

KARL LANDSTEINER (1868–1943):
The Austrian doctor Karl Landsteiner studied bacteria and **viruses** in the human body. After coming to New York to do research, he realized that humans have different blood types. His work led to matching blood types for safe transfusions. He won a Nobel Prize for this work in 1930. His early research with viruses helped scientist Jonas Salk to produce an anti-polio vaccine.

HANDS ON!

<u>FIND OUT YOUR COMPATIBLE BLOOD TYPES:</u> All you need is a piece of paper and a pencil. You can't donate blood until you are in high school and have a high enough body weight. But you can find out what blood types you are compatible with. First, find out what blood type you are. Then make a list of what types your blood type is compatible with. Type A blood can give to AB's, A's, and O's. Type B's can give to AB's, B's, and O's. O's can *only* give to O's. AB's can give to AB's, A's, B's, and O's.

38 Blood is made of billions of blood cells and a fluid called plasma.

Just one drop of blood contains millions of blood cells. But not all blood cells are alike. There are three different types of blood cells. Each type does a different job. Red blood cells carry oxygen and carbon dioxide. White blood cells attach themselves to and kill germs. The third kind are called platelets, which help your blood clot.

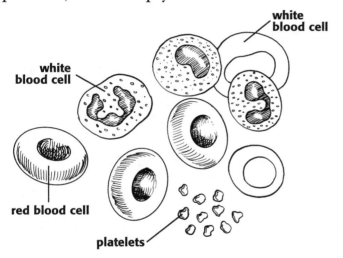

white blood cell

white blood cell

red blood cell

platelets

BONE MARROW AND BLOOD CELLS

Bone marrow is a squishy substance that is in the center of certain types of bones. Marrow makes blood cells and platelets. Red blood cells carry oxygen through the blood. White blood cells fight off disease. Platelets help heal wounds.

THE COLOR OF BLOOD

Blood is red because of the iron carried in red blood cells. When iron mixes with enough oxygen, it is a rust color. When it mixes with a lot of oxygen, it is red. As blood moves throughout your body and loses oxygen, it changes color and becomes darker and bluish. If all of the oxygen were removed (which never happens), your blood would turn blue! When oxygen is added from the lungs, it becomes red again.

ANEMIA

Red blood cells have the job of carrying and delivering oxygen to all parts of your body. **Anemia** is a lack of or dysfunction of red blood cells. With fewer red blood cells, an anemic person gets too little oxygen. Anemic people are often tired because of this limited oxygen supply.

Clotting is important when you injure yourself. If blood couldn't clot, it would just pour out of your body until it was all gone. All of these blood cells are in plasma, a liquid that makes up 55 percent of your blood. Plasma is mostly water, but it also contains salt, **vitamins,** and proteins.

HEMOPHILIA

Hemophilia is a rare hereditary disease that afflicts mostly men. A person with hemophilia can bleed to death from a small injury because his or her blood does not clot properly. People with hemophilia used to die at a young age, but now doctors can inject the clotting factor into a hemophiliac's blood. The shots are expensive and are needed every year.

Your blood plays an important role in healing injuries, like cuts and bruises.

When you cut yourself and start to bleed, almost immediately a scab starts to form. If it weren't for the scab, you would bleed to death. Tiny blood cells called platelets are part of your blood. Usually they flow along with the rest of the blood through the vessels. But if platelets run into something rough, like a cut in a vessel, they stick to the edge. They pile up and clump. The clump stops the bleeding, but your cut is still filled with blood. The blood forms a clot and changes from a liquid to a thick jelly. When the clot dries, the scab is formed. The scab protects the cut.

BURNS

A burn heals the same way a cut does. Usually a blister instead of a scab will form over a burn to protect it. Serious burns are dangerous because the skin cannot grow over the burned area fast enough. When this happens, germs can get in and cause infection.

Underneath the scab, white blood cells are doing their job of fighting off germs. Different types of white blood cells come in to remove old fibers, bits of damaged cells, and other white cells. Then, skin cells start to form over the cut to form a new layer of skin.

A bruise is an injury to your body. But unlike a cut, the skin doesn't break. When hit hard enough, blood vessels right under the skin break.

The blood flows out closer to the skin, which is what you see as the coloring of a bruise. The bruise changes color over time because the

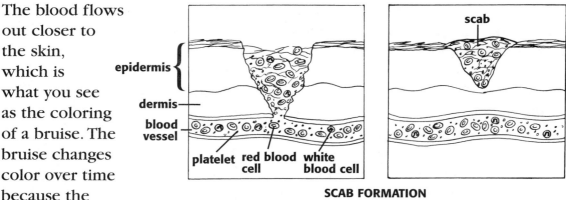

SCAB FORMATION

oxygen-rich blood cells, which make the area red, start to decay, turning it purple. The bruise disappears when the white cells are finished cleaning up.

40 Your immune system is like an army that protects you from diseases.

The king of fighters for your body is the immune system. Your immune system responds to unwelcome microorganisms. Your skin is the first wall of protection against invaders. Your body also produces many fluids to protect you. For instance, **mucus**—in your **sinuses,** eyes, and nose—kills bacteria. But if the microorganisms find their way into your body, there is another line of attack: **antibodies**.

Antibodies attack viruses, fungi, **parasites,** or other foreign substances. Specific antibodies are made to fight specific invaders. Antibodies are made

SNEEZING

A sneeze is a reaction of your immune system. When you sneeze, your body is trying to protect you from harmful dust or irritants by expelling them. Right before a sneeze happens, you may notice a certain itch in your nose. This itchy feeling is caused by pollen or dust that irritates the membranes in your nose. The membranes react and send signals to your brain. The brain then sends signals to the respiratory muscles. You are forced to inhale deeply and then sneeze. A sneeze can send mucus traveling up to 12 feet!

by the immune system's **B cells**. B cells are one of several types of white blood cells that make up the immune system. Each type of white blood cell does a different job:

B cell

- B cells are made in the bone marrow and go straight from the marrow to the blood. Once in the blood, B cells are on patrol looking for invaders and waiting to get messages from **T cells**. B cells have antibodies on them. If they run into an invader, they start splitting so that the army of fighters is larger. B cells work together to smother the invader.

T cell

- T cells are made in the bone marrow but spend time in the **thymus gland** before going out into the blood. Different types of T cells take charge once an invader has been found. Killer T cells multiply when they find an invader. Killer T's attach themselves to the invader and release chemicals that will kill it. Helper T cells send chemical messages to the B cells, telling them to start multiplying to attack an invader. Suppressor T cells tell the B cells when to stop fighting.

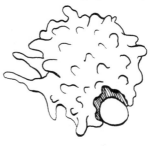

phagocyte

- **Phagocytes** are "eating" cells. When phagocytes meet an invader, they destroy it with chemicals and eat it. Phagocytes also clean up after B cells and T cells.

IMMUNE SYSTEM FAILURE

When the immune system fails, it is serious—even deadly. Some viruses, like smallpox, move through the body so quickly that the immune system doesn't have a chance to fight back. Another example of the immune system failing is in congenital immune deficiency, when the bone marrow of a fetus does not produce white blood cells to fight infection. Shortly after birth, the child usually dies. Autoimmunity happens when the body's immune system mistakes its own cells as invaders. It then destroys healthy cells.

41 The lymphatic system creates antibodies to fight invaders.

In addition to your immune system, your lymphatic system helps protect the body. Often called the second circulatory system, the lymphatic system is a web of vessels running throughout the body that interconnects with **lymph nodes**. The nodes filter out disease-causing organisms. The nodes are little pea-sized structures. They are located all over your body and in clusters around your neck, armpits, and groin area. The lymph nodes help make white blood cells and antibodies to help your immune system do its job. Lymphocytes are white blood cells created in the lymphatic system. They are some of the longest-living cells in the body, lasting 6–7 years.

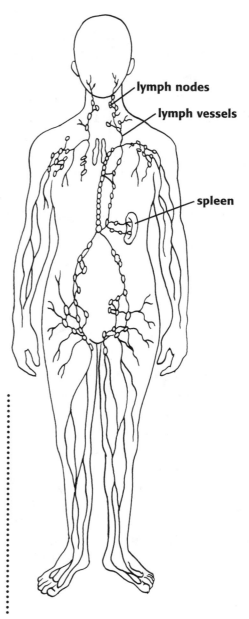

THE SPLEEN

The spleen is a structure that is part of the lymphatic system. It produces white blood cells and removes old red blood cells and other wastes from the blood. If you ever have an injury that causes you to bleed a lot, the spleen will provide the needed blood. Although the spleen is an organ that does a lot for your body, if it becomes injured, it is removed. An injured spleen can cause you to lose a lot of blood, and it cannot be repaired.

42 Diseases can be classified as either infectious or noninfectious.

Infectious diseases are caused by germs and viruses, caught from another person or from something in the environment. For instance, insects can pass on infectious diseases. Some infectious diseases you may have heard of are measles, the common cold, and tuberculosis. You may have heard

FUN FACT!

The most common noninfectious diseases occur in the mouth— for example, gingivitis (swollen gums).

the words "**contagious disease**." Contagious diseases are infectious diseases that are passed from person to person. Therefore, a disease like malaria, which is carried by mosquitoes, is not contagious. But the flu, which is spread by human contact, is contagious.

A **noninfectious disease** is inherited or comes from within the body. Examples of noninfectious disease are diabetes and anemia. Noninfectious diseases cannot be caught.

BRIEF Bio

ELIZABETH BLACKWELL (1821–1910): Elizabeth Blackwell was the first female doctor in the United States. She was rejected by 29 medical schools because she was a woman, but was finally admitted to Geneva College in New York. She had a hard time in medical school. As the only woman in her school studying to be a doctor, she was not taken seriously. After graduating, then practicing for a few years, she and her sister started a hospital in New York that eventually grew into a medical college and nursing school that accepted women.

VACCINATIONS

A vaccination exposes you to a certain infectious disease on purpose. When you get a vaccination shot or drink, you are taking in part of a disease-causing organism, or a dead one. The shot stimulates your immune system to produce cells that protect you from the disease caused by the germ. Your body responds as if it is being attacked, by making antibodies. Then, when you are exposed to the disease, your body already has the antibodies in place to fight.

43 Cancer is a noninfectious disease.

Doctors often say that cancer is not one disease, but many. This is because cancer can occur in so many places throughout the body—in tissue, glands, organs, skin, bone, blood, even muscle.

Cells in a healthy body keep a certain balance. The body makes just enough new cells to replace dead ones. Cancer happens when too many new cells are produced and cell growth is uncontrolled. The abnormal, fast-growing cells form a mass called a tumor.

It is believed that some types of cancer, like breast cancer, can be inherited. Many scientists believe that you may inherit a tendency to develop cancer. Then outside factors will set it off—for example, lung cancer caused by smoking. Cancer has also been found in people regularly exposed to chemicals such as asbestos, an insulator that is fire resistant.

Many types of cancer can be cured, depending on where the cancer is and how long it has been growing. Radiation and chemotherapy are methods doctors use to try to kill cancer cells in cancer patients.

44 Bacteria are one-celled organisms.

We often talk about germs being on a glass that someone else has used, or in a dirty area. But what exactly are germs? Germs are bacteria that can cause infection. Bacteria are one-celled living things, so small that you can only see them under a microscope. They are all over our bodies and everywhere else—in water, food, soil— you name it! But not all bacteria are bad; in fact, some of them help us. Germs, however, are disease-causing bacteria.

Where do you pick up germs? You can catch them from other people

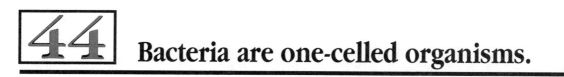

**E. coli,
a common bacterium found
in the human intestines**

who are infected with them. Or you can get them from food that is not properly stored, like milk that has been left out in the sun for a long time. A healthy immune system fights unwanted bacteria, but sometimes they sneak in and make you sick before your immune system can fight them off.

For illnesses caused by bacteria, your doctor can figure out which bacterium has attacked your body and prescribe an antibiotic medicine. Antibiotics kill bacteria by attacking them or interfering with their ability to divide.

BRIEF Bio

SIR ALEXANDER FLEMING (1881–1955): Fleming was a Scottish bacteriologist (a scientist who studies bacteria). In 1928, he was studying the influenza (flu) virus. He accidentally grew some mold on one of his culture plates. He found that the mold, *Penicillium notatum,* had created a bacteria-free circle around itself. This mold led Fleming to develop the medicine we use today, penicillin. After another scientist, Dr. Howard Florey (1898–1968), tested it, and Dr. Ernst Chain (1906–1979) managed the construction of a plant to make it, penicillin became available to the public. Today, we use penicillin as an antibiotic to kill bacteria. Fleming, Florey, and Chain won the Nobel Prize for their work with penicillin.

PARASITES

Parasites are organisms that live off of your body and can make you sick. They are larger than bacteria and viruses. You've probably heard of lice, tiny creatures that live mainly in human hair. Lice are parasites that feed on blood. Scabies are tiny mites that live in the outer layer of the skin. Worms that can infect the digestive system include roundworms, pinworms, and threadworms. These worms live in the stomach and can cause itching around the anus. Tapeworms are long and live in the intestines.

45 A virus lives in cells.

Viruses are even smaller than bacteria, and they operate a little differently. Viruses are usually looking for a host, or a place to live, like the cells in a body. A virus won't live just anywhere. For instance, the chickenpox virus will live only in humans. Once the virus gets into the type of body it needs, it floats around in the bloodstream looking for the particular cell in which to reside. When it finds the cell, it moves in and changes the chemistry of the cell. The cell is tricked by the virus and starts following the orders the virus gives. Then, instead of splitting and reproducing itself, the cell will start reproducing the virus within it. The cell fills up with these viruses and then bursts, letting all of the newly made viruses out into the bloodstream. They go looking for homes and start the process all over again.

CHICKENPOX

Chickenpox, a disease caused by a virus, is common in children. Chickenpox produces blistery bumps all over the body, as well as a fever. There is no cure, but you can only get chickenpox once. Once your immune system learns how to fight it, you will never get it again.

The problem with trying to use medicine to fight a virus is that the virus is living inside the cell and is protected by it. Therefore, if a medicine is able to kill a virus, chances are it would kill the cell, too.

NO CURE FOR THE COMMON COLD

Both the common cold and the flu are infectious diseases caused by viruses. There are about 200 different types of cold viruses you could catch. The flu is caused by influenza A or B virus. Both a cold and a flu infect the upper respiratory system, but a flu virus may spread to the lungs. A flu causes more uncomfortable symptoms than a cold, such as the chills, a fever, and body aches. You don't get a fever with a cold.

46 AIDS is caused by HIV, a virus that attacks your immune system.

AIDS stands for acquired immunodeficiency syndrome. It is a disease that develops in people who have been infected by a virus called HIV, human immunodeficiency virus. HIV can live quietly in the human body for several years. The virus attacks white blood cells. When the supply of white blood cells is depleted and the immune system starts to fail, the HIV infection becomes AIDS. Because of a weak or destroyed immune system, people who suffer from AIDS are easily infected by diseases that otherwise would not be as harmful.

HIV cannot be spread by kissing or sharing a glass. HIV can only live in blood and body fluids. Therefore, the only way to spread it is by a transfer of blood or body fluids. Transfer can take place during sexual activity, or through a break or cut in the skin. Transfer can also take place between drug abusers who share needles. So far, there is no cure for AIDS.

HISTORICAL DISEASES

1. From 1347 to 1351, the Black Death was an infectious disease that killed anyone who caught it.
2. Smallpox ripped through communities 200 years ago.
3. In 1918 and 1919, the Spanish flu infected 20 percent of the population, killing 20 million people.
4. Polio, a disease of the 1940s, struck suddenly and caused sufferers to be crippled or to die.

 **Your nervous system consists of the brain, the spinal cord, and nerves.**

The nervous system controls every movement you make. This includes voluntary movements, like turning on the television and finding a channel you like. It also includes involuntary movements, like digesting food and the pumping of your heart. The nervous system also lets you feel things, such as pain and temperature. All of these movements and feelings happen because of messages the brain sends out and receives.

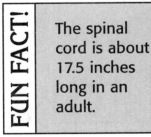

FUN FACT! The spinal cord is about 17.5 inches long in an adult.

Your brain is the command center of the nervous system. The spinal cord is like a thick wire that delivers commands between your body and your brain. It attaches to the base of your brain and runs down your back. The main nerves that branch off the spinal cord are spinal nerves. They are like "wires" that pass through gaps in the backbone. There are 31 pairs of spinal nerves. The spinal nerves connect to smaller "wires" so that nerves can reach all parts of the body. Consider this the next time you touch something warm with your hands. A message from your fingers will travel through the nerves in your arms to the spinal cord and up to your brain.

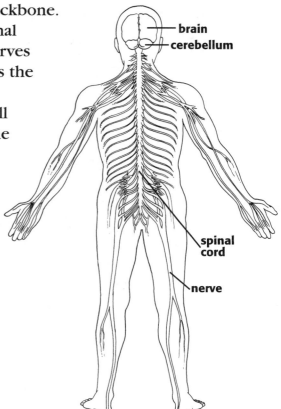

PARALYSIS

Paralysis is the loss of control over certain parts of the body. It results from damage to the spinal cord that blocks the communication between the body and the brain. If damage is done to the middle or lower spinal cord, a person may become a paraplegic. Paraplegics do not have control over their legs and sometimes do not have control over their torsos. Damage done to the spinal cord in the lower neck area can cause quadriplegia, meaning the whole body from the neck down is paralyzed.

HANDS ON!

MAKE YOUR ARMS MAGICALLY RISE: Stand in a doorway and push your arms out so the tops of your hands push against the sides of the doorway. Do this for 1 minute. Then step out of the doorway. What happens? Your arms will rise automatically. Why? Because for a minute or so, your brain told your arms to lift up. When you step out of the doorway, some of the commands are still on their way to your arms. When the commands reach your arms, they lift up.

48 Neurons are the nerve cells of your nervous system.

Neurons are special cells of the brain, spinal cord, and nerves. A neuron is microscopic, but if you could see one, you would notice that it looks sort of like an octopus with a long tail.

A neuron has three main parts. The cell body is the central part. Extending off the body are short branching arms called **dendrites**. The dendrites receive incoming messages. The third part is called an **axon,** which carries outgoing messages. An axon is like a long string. Axons connect the neuron with other neurons or organs, such as muscles. If a message travels from a neuron into a muscle, it makes the muscle contract. Some axons, like those connecting nerves in your feet with the spinal cord, can be up to 3 feet long!

FUN FACT!

A neuron can be attached to 50,000 or more other neurons! The brain is made up of 100 billion neurons.

BRIEF Bio

LUIGI GALVANI (1737–1798): Galvani was an Italian scientist who performed experiments with electricity. He discovered that the muscles in dead frog legs would contract when touched with an electrical spark. This led him to believe that our movements were powered by some sort of electrical impulse. He became known as "The Frog's Dancing Master."

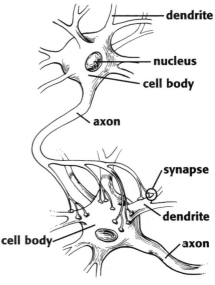

dendrite

nucleus

cell body

axon

synapse

dendrite

cell body

axon

ELECTRICAL AND CHEMICAL NEURAL POWER

How do neural messages travel? They travel by tiny bursts of electrical power—a fraction of the power a flashlight uses. The electrical power helps the message travel along the length of a neuron. At the place where neurons meet is a small space called a **synapse**. A message "jumps" over the synapse with the help of chemicals called **neurotransmitters**. Once it gets to the next neuron, the message is again powered by an electrical impulse.

HANDS ON!

CATCH IT IF YOU CAN: Are your neurons delivering messages quickly enough? Cut out a piece of paper about 4 by 5 inches. Drop it from over your head, then try to catch it. Try it with both left and right hands.

A **reflex** is an emergency reaction.

Touching something really hot causes an instant reaction—you pull your hand away quickly without even thinking about it. Why? When the nerve endings in your fingers touch something dangerous, like a hot stove, the nerve impulses don't take their usual path to the brain. The emergency message makes its way to the spinal cord and splits. Part of it loops back to the muscle where the pain is and makes you pull back or react. The rest of the message still makes its way to the brain, but in the meantime, your reflex has already caused you to react.

Another type of reflex is checked when you get a physical exam. The

HANDS ON!

FIND YOUR OWN REFLEX: Take turns doing this with a friend. Have your friend sit on something high enough so that his or her legs dangle. Tap right below your friend's kneecap with the side of your hand. You don't have to tap very hard, but you may have to tap around to find the reflex. You'll know when you find it. Your friend's foot will jerk up. Switch places, and have your friend find your reflex.

doctor will tap your knee, and your leg moves without you giving the command. The message sent to the spinal cord loops back to tell the muscle in your leg to contract. This is how your doctor checks the connections to your spinal cord.

50 The **cerebral hemispheres, cerebellum,** and **brain stem** are the main parts of the brain.

The brain is the control center of the body. It is here that we think, imagine, store memories, and have emotions. The brain is an organ made up of 100 billion neurons! Slightly larger than a grapefruit and weighing about 2.75 pounds, the brain looks like a big wrinkled walnut. Although often spoken of as gray matter, the brain is gray and white.

Different parts of the brain do different things:

- The cerebral hemispheres make up 80 percent of the brain. This part of the brain is where your thinking and learning takes place.
- The cerebellum, in the middle of the brain, monitors all the movements you make with your muscles, from drinking to running. It also helps you stay balanced.
- The brain stem connects the brain and the spinal cord. It controls body functions needed for survival, such as breathing. The **thalamus** in the brain stem organizes all information coming into the brain.

FUN FACT!

The brain receives 100 million messages every second from all over the body.

BRAIN MEMBRANES

Both the brain and the spinal cord have fluid around them. They also have three layers of **meninges,** or protective membranes. You may have heard of a disease called meningitis. Meningitis develops when germs infect the fluid and the meninges. The meninges can swell, which can cause damage to the brain or spine. In small children, meningitis can be fatal.

The **hypothalamus,** also in the brain stem, controls body temperature, heart rate, blood pressure, and hormone releases, as well as memory and sensory experiences.

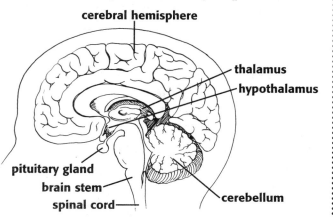

cerebral hemisphere

thalamus

hypothalamus

pituitary gland

brain stem

spinal cord

cerebellum

DYSLEXIA

Dyslexia causes a person to mix up the order of letters, words, symbols, or directions. Although it may be caused by vision problems, doctors think it could also have to do with a brain disorder. Dyslexia can affect anyone, regardless of **intelligence**. Therapy can help people with dyslexia learn how to improve their reading and writing abilities.

ALCOHOL AND THE BODY

For most adults, an occasional drink is OK. When a person drinks an alcoholic beverage, the alcohol goes into the blood. Since blood travels everywhere in the body, so does the alcohol. Even with just one drink, a person does not have 100 percent control over his or her behavior, coordination, and thinking. With three or more drinks in an hour, a person is usually legally intoxicated. This means that the person should not drive a car or do anything else that requires clear thinking.

51 Damage or changes in the brain, spinal cord, or nerves can cause disorders.

Many disorders can develop because of some sort of damage or change in the nervous system. Disorders can cause weaknesses, paralysis, or seizures.

■ One person in 200 suffers from epileptic seizures. These seizures are a result of the brain having uncontrollable electrical activity that may cause a person to have involuntary movements. Epilepsy can happen in children and adults. Doctors can't always tell what causes epilepsy, but in adults, it is sometimes linked to a head injury, stroke, tumor, or some other brain condition.

■ Multiple sclerosis (MS) affects one person in 1,000. This disease attacks the protective coating that protects nerve cells (called myelin sheaths). Nerve cells without this protection have difficulty delivering messages to one another. MS can cause episodes of clumsiness, blurred eyesight, even paralysis. The episodes can last a few weeks. As the disease progresses, permanent damage to the nerves is done. There is no cure, but treatment to slow down the episodes is given to some patients.

■ Parkinson's disease occurs in about one in every 200 people over the age of 60. It is caused by too little dopamine, a neurotransmitter, in the brain. Parkinson's disease causes muscles to be weak and stiff. A person with Parkinson's often has hands that tremble.

■ Alzheimer's disease occurs often among the elderly. The brain produces too much of a particular protein, which causes brain damage. The damage causes the sufferer to forget things. At first a person with Alzheimer's disease will forget small things, but eventually he or she will forget everything, including how to cook and drive, or even recognize family and friends.

52 | The hypothalamus controls body temperature.

Have you ever wondered how your body keeps you warm? You know to wear warm clothes when it's cold out, but it takes more than just a wool sweater. Humans and other warm-blooded animals have a built-in thermometer. Your house may have a thermometer that tells the heater to come on if the air gets too cold. When the temperature is just right, the thermometer tells the heater to shut off. The built-in thermometer in your body does the same thing. It's called the hypothalamus.

The hypothalamus can tell the temperature of your body by the temperature of your blood. When your body gets too hot or too cold, the hypothalamus sends messages to the body. When you're too cold, for instance, you start to shiver. Shivering makes your muscles move more, and this helps keep your body warm.

Your skin helps control your temperature, too. If you're cold, your body sends less blood to your skin in order to conserve heat. If you're hot, your skin sweats in order to cool off your body.

53 The left brain and right brain control opposite sides of your body.

The brain has a distinct left side and a distinct right side. The left side of your brain controls the right side of your body. And the right side of your brain controls the left side of your body. In most people, the left side specializes in reasoning and things like language, math, and writing. The right side is the creative side and specializes in art, music, and creative

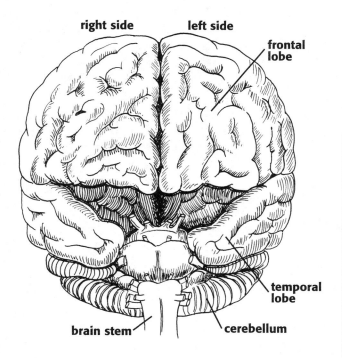

right side   left side

frontal lobe

temporal lobe

brain stem   cerebellum

HANDS ON!

DETERMINE YOUR DOMINANT EYE: Extend your arm out in front of you with your thumb pointed up. Close one eye. Now switch and close the other eye. Go back and forth a few times. You'll see that your thumb appears to jump toward one side. If it jumps to the right, your right eye is dominant. If it jumps to the left, it means your left eye is dominant.

thinking. However, the two sides work together. They communicate through bundles of nerve fibers.

In almost everyone, one side of the brain becomes dominant. For most people, the left side is dominant. These people are right-handed. You probably already know which hand you always use to write with, but which foot do you most often use to kick a ball? Which arm do you use more in sports? Try to notice these things next time you do something.

54 There are two kinds of memory: short-term memory and long-term memory.

How do things like a friend's phone number or a teacher's name become part of your memory? Short-term memory lasts about a minute and can store only five to seven items at a time. Long-term memory on the other hand, lasts longer, even for years, and it can store trillions of pieces of information. When you first learn about something, it may be stored in your short-term memory. For instance, you may remember a phone number for just a minute and then forget it. But if you keep calling the same phone number over and over, it becomes part of your long-term memory, where it is stored.

Do you ever meet someone and then forget his or her name right away? You may remember the face the next time you meet, but not the name. Many people can remember visual pictures better than words or numbers. Experiments show that memory takes in pictures and objects you see more directly than it takes in words and numbers.

HANDS ON!

USE MNEMONICS TO HELP YOU REMEMBER:
Mnemonics involves making up a word, sentence, or story in order to remember something else. For example, to remember the parts of the digestive system—Mouth, Esophagus, Stomach, Small intestine, and Large intestine—you might think of the sentence, My Ear Seems So Little, using the same first letters of the names of the digestive organs. Try it next time you have to memorize something for homework.

BRIEF Bio

IVAN PAVLOV (1849–1936): If you have a pet dog or cat, you may notice that just the sound of opening a can or pouring dry food into a bowl will bring the animal running. This is the sort of conditioning that Pavlov, a Russian physiologist, studied. He experimented with a dog. He would regularly ring a bell right before giving meat to the dog. After doing this over and over, the dog soon linked the sound of a bell with eating meat. Each time the dog heard the bell, he expected to eat meat. This conditioning explains why your pet hears a certain sound and links it to eating.

Pavlov's research helps explain why we have certain preferences and dislikes. If, for instance, a certain song was always on the radio during a pleasant time in your life, you may enjoy hearing that song and not be fully aware of the links that make you like it so much.

REMEMBERING NOTHING

Often on television or in the movies, we see characters who cannot remember who they are. They have amnesia. In real life, this drastic form of amnesia is rare. Usually, a blow to the head, if hard enough, will cause a person to lose memories of the last hour or so. These are memories that were in short-term memory and had not yet moved into long-term memory. If the injury damages a certain brain area called the hippocampus, a person cannot create new memories. This means the person would have all of his or her old memories intact, but each day he or she would wake up and only have memories leading up to the time of the head injury.

55 Your intelligence is partly inherited and partly from your environment.

How can we tell how intelligent or smart someone is? We can't, exactly. In fact, today scientists still have not come up with one meaning for the word *intelligence*. People consider many different factors when talking about intelligence, such as speaking skills and math skills, one's ability to figure things out on one's own, and creativity. Many IQ (intelligence quotient) tests have been created to measure how smart a person is, but doing poorly or well on the test still does not show how smart you may be in areas not covered on the test.

Scientists look at two major areas when considering where intelligence comes from:

■ Heredity factors. The proof that intelligence is partly hereditary comes from IQ tests. When identical twins took IQ tests, their scores were very close. Identical twins have exactly the same **genes,** which are passed down from the parents. When sisters and brothers took the tests, their scores were also close, but not as close as the identical twins. How much of one's intelligence is inherited is still in question today.

■ Environmental factors. Some tests show that surroundings and training help determine intelligence. Current research indicates that babies who are stimulated as infants—by looking at art or listening to classical music—are better in their development and thinking ability. Another part of environment is education. Someone who experiences an enriched education may have a higher intelligence than someone who is uneducated.

HANDS ON!

SOLVE THE PROBLEM:
Intelligence tests focus on your knack for solving problems. Can you solve the problem below?
Say you go to a friend's house. When you get there, your friend is not at home, but you find a note your friend has left. The note is ripped into four pieces, each with four letters on it. Can you tell what the note says?

Paper #1: W C B S
Paper #2: I O A O
Paper #3: L M C O
Paper #4: L E K N

Even though there are many different personality types, each one of us is unique.

56

What kind of person are you? Do you keep your room very neat, or are you messy? Do you finish your homework on time or ahead of time? Are you a

ANSWER: The note said: WILL COME BACK SOON

leader or a follower? Are you outgoing, or quiet? Many details about how we behave, our emotions, and what we like or dislike make up our personality.

There are many tests you can take as an adult that will tell you what type of personality you have. But even with the results of a test, personalities are complicated. Even if two people are very organized, one may react to stress very well, while the other person may not be able to handle stress. With all of the combinations of personality traits, each person ends up having his or her own unique personality.

Personality traits—certain ways that you are—have both heredity and environmental origins. For example, a person who is an only child may have different ways of doing things than a person who comes from a family of six. He or she has grown up in a different type of environment. A person who is very tall may behave differently than a person who is very short. This could be because he or she is treated differently by the public.

PHINEAS GAGE

In 1848, Phineas Gage was working in Vermont blasting rock for a railroad line. He accidentally pounded some explosives with an iron pole, causing an explosion that sent the pole through his head! The pole went under his left eye, through and out the top of his skull. He recovered completely, but something strange happened. Somehow this accident caused Gage to have a personality change. Before the accident, he was friendly and a generally nice person. After the accident, he was impatient and unfriendly. This proved to brain specialists that the part of the brain that was damaged in Gage controlled emotions and thoughts. Because of the brain damage, Gage changed the way he expressed himself.

57 Your emotions are connected to your body.

When you have a certain emotion, it is not just a feeling in your head— your whole body has a physical reaction, too. Think about what happens when you become angry or afraid. Your heart might pound faster, and you might feel the release of the hormone adrenaline, giving you a burst of energy. Even though the emotion is anger or fear, the energy helps you deal with that anger or fear. The energy might help you run away from a situation that is dangerous, or fight back in a situation you need to confront.

Think about when you are happy. How does your body react to happiness? What about when you are sad or upset? You may get a stomachache or a headache when you're upset. One thing to remember is that everyone handles emotions differently. Some people can get very angry and never show it. Others are short-tempered and their anger may flare up very easily. Next time you have a strong emotion, try to pinpoint how your body reacts to that emotion.

LAUGHING AND CRYING

Laughing and crying both happen because of strong emotions. If you think about it, your body makes similar movements with both. You take in more air and let out big gulps. Your shoulders may shake when laughing or crying. Both can also be a type of relief. If someone is overwhelmed with emotions, he or she can even go from laughing to crying without being able to help it. Patients who have suffered brain damage have given doctors proof that many of the same brain circuits and muscles move during both laughing and crying.

58 Mental health is the healthy state of your emotions.

It is easy to know when you are physically healthy: you eat a balanced diet, get rest and exercise, and feel good. If a person is unhealthy, it may be a result of not eating well or not resting. An unhealthy person may get sick a lot. But what about mental health?

Mental health is the health of your emotions. Everyone experiences a range of emotions—from an energetic happiness all the way down to depression. When do emotions become unhealthy? It is normal to experience all feelings, including sadness. Certain events, like death, can cause a person to feel really sad or depressed for a long

MANIC DEPRESSION AND SCHIZOPHRENIA

Manic depression is an illness in which a person goes from having a high-energy phase to a low of depression without a reason. It can be frustrating, because that person has no control over these extreme highs and lows. During the low phase, the person does not have a lot of energy to get things done.

A person with schizophrenia lives in a fantasy world. He or she may have false thoughts. Many schizophrenics have the false thought that someone is after them. They also see imaginary things and hear voices. Many mental illnesses, like manic depression and schizophrenia, are caused by an imbalance of neurotransmitters in the brain. For instance, it is believed that schizophrenics have too much of a certain kind of neurotransmitter. Both illnesses are at least partially treatable.

DRUG ABUSE AND EMOTIONAL HEALTH

Drugs are terrible for you! They're bad for your body and can cause your emotions to go haywire. Drug users can become withdrawn, overactive, depressed, anxious, or irritable— depending on the drug. Even marijuana, which is often called a harmless drug, can cause memory loss, a loss of judgment, and slowed movement. Not only do drugs alter your thinking and ability to do things while you take them, but if you decide to stop, the problems don't necessarily go away. The same is true of alcohol. Alcohol abusers may quit drinking, but still have permanent damage to their memory and thinking skills.

time. Feeling sad is a way for someone to heal and get over a loss. Anxiety is another feeling that is a normal response to stress. You may feel anxious before a big test, for instance. But when a person feels depressed or anxious all the time, it could be a sign of a mental health problem.

One third of your life is spent sleeping.

Have you ever tried to stay up really late, or even all night? It's hard to do. If you succeed, the next day your body shows signs of the lack of sleep. You may be cranky or have trouble paying attention. If

> **FUN FACT!**
>
> A person who weighs 150 pounds burns about 1 calorie per minute while sleeping. Multiply that by how many minutes there are in a night!

SLEEPING DISORDERS

Insomnia and narcolepsy are two kinds of **sleeping disorders**. Insomnia is the inability to sleep night after night. Sometimes noise is to blame. Insomnia is usually caused by nervousness or depression. People who have narcolepsy fall asleep all the time. This can happen in the middle of a party or a class. Some narcoleptics even fall asleep standing up or when they are walking!

When you sleep, your body relaxes, but your brain doesn't. In fact, your brain is busy the whole time. Scientists know this because they can measure brain waves. They have learned through experiments that during sleep, the brain has two general stages. When you dream, you are in **REM (rapid eye movement) sleep**. During this time, the brain is as active as it is during the day. The other sleep stage is slow-wave, or non-REM, sleep. During this time, your brain waves slow down and your body temperature and pulse rate drop. There are four stages of non-REM sleep. Throughout the night, you go back and forth between REM and non-REM sleep in a repeating cycle.

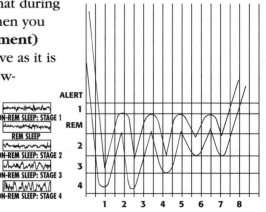

you tried to stay awake for 48 hours (two days straight), it would be nearly impossible. During sleep, the body restores energy for the next day. Not everyone needs the same amount of sleep. Adults need an average of 8 hours of sleep each night, but babies sleep for 16–18 hours a day. And some adults can get by on only 5 or 6 hours. Young children need 8–10 hours a night.

60 Dreams help you "digest" the day's thoughts.

Many people cannot remember their dreams, but experts say that everyone dreams. And they say it is important to dream. In dream studies, doctors woke up patients each time they started to dream (during REM sleep). The next day, even though the patients slept, they were very tired because they were not allowed to dream. Some researchers think that dreaming is necessary to make sense of what happens during the day. Others think your dreams can help you by showing you how to solve problems you encountered during the day.

BRIEF Bio

SIGMUND FREUD (1856–1939): Freud invented a scientific approach to studying human behavior. He thought that people's emotions were hidden and would be revealed through their dreams. He was famous for a book called *The Interpretation of Dreams*. In this book, he presents his theory that dreams are a way to express forbidden wishes.

HANDS ON!

<u>WRITE DOWN YOUR DREAMS:</u> **Keep a dream diary. Each morning right after you wake up, jot down the dreams that you remember. Try to pay attention to detail, and also write down how you remember feeling in the dream. After a week, look back at your notes. Do the dreams have anything to do with what's going on during the day? What do you think they mean?**

61 There are five senses: smell, taste, sight, hearing, and touch.

Your senses are the gateway from the brain to the outside world. How would your brain know what lay in front of you without your eyes? Or how would your brain be able to know what sounds surround you if it weren't for your ears?

BRIEF Bio

Helen Keller (1880–1968): At the age of 19 months, Helen Keller came down with a fever. The illness left her deaf and blind. But Keller overcame both deafness and blindness with the help of teacher Annie Sullivan. Sullivan taught Keller sign language. Keller remembers learning her first word. Someone was pouring water onto her hand and signing the word *water* into her hand. She suddenly remembered "water" from her early days before she became sick. That word opened the door for her. She soon learned how to write and speak. She wrote her autobiography when she was 23.

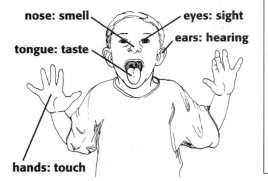

nose: smell
eyes: sight
ears: hearing
tongue: taste
hands: touch

MINUS ONE OR MORE OF THE SENSES

Many people do not have all five senses. A person without one or more of the senses could have been born that way. In other cases, a sense may be lost because of an accident or aging. Not having one of the senses does not mean that anything is wrong with a person. In fact, a person who does not have one of the senses must be very creative in order to adapt.

You may be familiar with sign language. Many people who are hearing impaired or deaf use sign language to communicate with others. A person who is visually impaired may not have full use of his or her eyesight, or may be blind. People who are visually impaired often have guide dogs or tapping canes to help them feel their way around. A language for the blind called Braille is a code that can be read with the use of fingertips. Look around for Braille signs. They are sometimes located in banks near the automatic teller machines (ATMs), and in or near an elevator.

Your senses help you navigate your way through each day. In the morning, you hear your alarm clock ring. You smell and taste your breakfast. You know the way to the bus stop by sight. And touch tells you how to dress for the day. All senses rely on the messages nerves send.

The senses also help protect you. In the case of a fire, you can hear a smoke alarm and you can smell or see the smoke. By touching a door or a wall, you can feel whether it's hot from the fire.

62 Your eyes work like little cameras taking pictures of the world.

Your eyes are like two cameras that take pictures and send them back to your brain. Once your brain receives an image from your eyes, you instantly understand what you're looking at.

Here's how it works. Light first enters through the **cornea,** the outer surface of the eyeball. It then passes through the **iris** and **pupil** to the **lens**. The lens focuses the light on the **retina** at the back of your eye. The retina sends nerve signals to the brain, which translates the signals into an image of the object you are looking at.

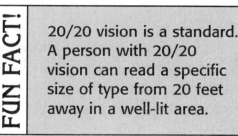

FUN FACT!

20/20 vision is a standard. A person with 20/20 vision can read a specific size of type from 20 feet away in a well-lit area.

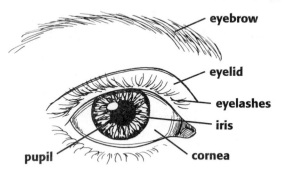

eyebrow
eyelid
eyelashes
iris
cornea
pupil

NEARSIGHTEDNESS AND FARSIGHTEDNESS

About 30 percent of Americans are nearsighted, and about 60 percent are farsighted. Nearsighted people can focus more easily on near objects than distant objects. Farsighted people can focus more easily on distant objects than near objects. In most cases, both nearsighted and farsighted people can correct their vision with glasses or contact lenses.

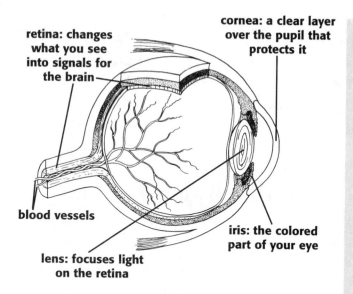

retina: changes
what you see
into signals for
the brain

cornea: a clear layer
over the pupil that
protects it

blood vessels

lens: focuses light
on the retina

iris: the colored
part of your eye

The retina is made of special cells called **rods** and **cones**. Cones enable you to see color and bright light. Rods let you see black and white and allow you to see well when there is dim light. If you turn out the light in a room, your eyes take a few minutes to adjust to the dark. They are switching from cones to rods.

Your two eyes work together to give you depth perception. Cover one eye and try to reach for something at arm's length away. Without the other eye, you might miss your target! We also have what is called peripheral vision. Even if your eyes are focused straight ahead, you can still see to both sides of your body.

63 Blinking is for moistening and protecting your eyeballs.

Have you ever tried to stop blinking? If you have, you probably couldn't do it for very long. Blinking is as natural as breathing. You blink every 2–10 seconds without even thinking about it. Like many other movements that happen automatically, blinking has a specific purpose—to protect your eyeballs. In order to move around freely, your eyeballs need to stay moist. The insides of your eyelids and the outsides of your eyeballs are covered with a transparent oily coating. Eyelids are like windshield wipers that coat the eye with moisture each time you blink. Eyelids also keep your eyes clean, brushing away particles that may be on the cornea. And they can block out lights that are too bright. Your eyelashes also shade your eyes and keep them clean.

FUN FACT!
You blink about 15,000 times in a day.

MOIST EYES

Your eyes are always moist. Underneath the oily outer coating on your eyeball is another layer of mucus that also helps keep your eyes moist. If your eyes become irritated, they will produce tears, which contain salty water. Tears protect your eyes by washing away irritants.

HANDS ON!

COUNT YOUR BLINKS: For a whole minute, count how many times you blink. Take that number and multiply it by 60. This number will tell you how many times you blink in an hour. Multiply that by 24, and you'll get the count of a day's worth of blinks. This activity works best if you count someone else's blinks, or have someone count your blinks.

64 Your nose smells, and filters and fights germs.

Other than adding character to your face, your nose has many purposes. Smelling is its most obvious function. But your nose protects you in many ways, too.

■ Smelling. Microscopic particles—**molecules**—float in the air. You inhale

FUN FACT!
Your nose has about 30 million cells to receive smells. Humans can distinguish about 10,000 different smells!

some of them through your nose. Way up high in your nose is a patch of wet mucus. Poking through the wet patch are nerves that lead to your brain. The molecules stick to the mucus, and the nerves send a message to the brain to tell you what you're smelling.

■ Filtering, and fighting germs. Hairs inside your nose act as a filter. You can see these hairs if you tilt your head while looking in the mirror. They catch dust, pollen, and other particles floating around. About every 20 minutes a new supply of mucus is produced in your nose. Tiny hairs in your nose sweep the mucus, and any particles from the hairs in the front, toward the back of your throat. You swallow it, and the stomach juices kill any germs in the mucus.

■ Temperature control. In the back of your nostrils, the air passes by blood vessels that warm it up before it goes to your lungs.

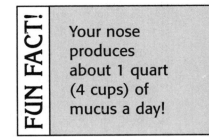

FUN FACT! Your nose produces about 1 quart (4 cups) of mucus a day!

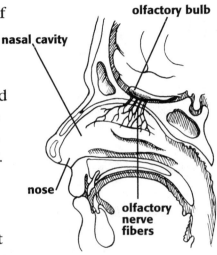

olfactory bulb

nasal cavity

nose

olfactory nerve fibers

HANDS ON!

DETERMINE THE LENGTH OF A SMELL: Spray some perfume, and time how long it takes for you to get used to the smell. Try some other smells, too, like hand cream. What happens? The nerves in your nose get tired when they take in a lot of one smell. This explains why when you smell a strong odor, after a few minutes, it seems like the smell goes away. But what really happens is that you just get used to the smell being there.

SAFETY ALERT
Be sure to ask your parents for permission to use the products in the smell activity. Do not spray anything toward your face.

SMELL AND MEMORY

You may not pay too much attention to smells in the air, unless you smell something really bad or really good. You might notice sometimes that you will get a whiff of a familiar smell that suddenly reminds you of a certain time, place, or person. Of the five senses, it is the sense of smell that is most powerful in reminding you of things from the past.

65 | Openings in your skull are called sinuses.

If you thought your ears, mouth, and nostrils were the only openings in your skull, think again. You have eight nasal sinuses, which are like air pockets in the front of your skull. Sinuses, like the nose, make mucus that, most of the time, drains down to your nose and throat. If you have ever slept on your side and woken up with one side of your head clogged, it was because your sinuses were not draining. Most of the time the sinuses are empty, except for the mucous lining that protects them. When your nose doesn't produce enough mucus, your sinuses will make up for it. If your sinuses are infected, they will fill up with pus and extra mucus. If you have allergies, they may become swollen, itchy, and runny. Your sinuses have a lot to do with how your voice sounds.

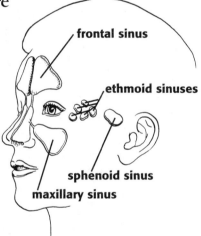

frontal sinus

ethmoid sinuses

sphenoid sinus

maxillary sinus

SINUS INFECTIONS

Sinus infections usually start after a cold or **allergy** attack. The sinuses, infected by bacteria, cannot drain properly and become clogged. Sinus infections can cause terrible headaches, but they can be treated with medicine.

66 Your tongue is covered with 10,000 taste buds that allow you to taste.

Next time you sit down to eat, hold your nose. Can you still taste your food? Probably not—or at least not very well. The same thing happens when you have a cold. You cannot taste very well. This is because the sense of taste relies on the sense of smell. Smell and taste work together. Smell tells you what flavor something is while your tongue can detect four basic tastes—sweet, sour, bitter, and salty.

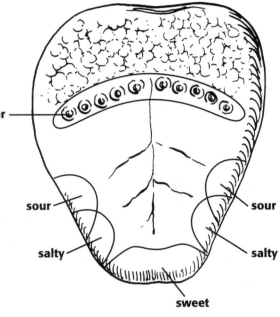

Your tongue, both a muscle and an organ, is covered with bumps. Each bump has sensory cells in it, which are the taste buds. Taste buds detect specific tastes; some can taste sweetness, while others are sensitive to sourness.

HANDS ON!

TAKE A TASTE TEST: This activity is fun to do with a friend. Blindfold a friend so he or she can't see the test samples. Then have your friend hold his or her nose and taste two different things, like a slice of apple and a slice of onion, or a slice of orange and a slice of lemon. Try to give your friend two foods that have the same texture but have really different tastes. Can your friend tell the difference? What does this tell you about how your nose helps you taste?

67 Your ears capture vibrations and allow you to hear sounds.

If you look at how your ears are shaped, you can see that they are curved in order to capture sounds. There are three main parts to the ear:

- The outer ear is shaped especially to capture sounds and funnel them into your middle ear.

middle ear

anvil

hammer   stirrup

semicircular canals

outer ear

inner ear

cochlea

outer ear canal

eardrum

middle ear

BRIEF Bio

LUDWIG VAN BEETHOVEN (1770–1827): Beethoven, a famous composer, grew to be deaf over the course of his musical career. He began his musical training at a very early age. He started to perform when he was 12 and was a well-known musician by his 20s. At the age of 30, he started to lose his hearing, but he continued to compose. Having had lots of experience with musical sounds, he could still hear the music using his imagination. He composed some of his greatest symphonies when he was completely deaf.

HANDS ON!

HEAR THE VIBRATIONS OF SOUND: Sound is really a vibration. Take a rubber band and wrap it around a mug, with the band going across the opening. Pluck the rubber band and watch it. You will hear sound as long as it is moving, or vibrating. When the rubber band stops, the sound stops. All sounds work this way, but usually the vibrations are too small to see.

- In the middle ear, the **eardrum** is a thin piece of tissue through which sounds—vibrations—pass. The eardrum transfers the vibrations to a mini-amplifier made of tiny bones called the hammer, anvil, and stirrup.
- The inner ear contains the cochlea, a spiral tube filled with fluid that sends signals to a nerve that tells the brain what you are hearing.

 ## Your inner ear helps you keep your balance.

The inner ear helps you stay balanced. Even when you close your eyes, you can stand up straight without falling down. This is because of a system of three semicircular canals in the inner ear. The canals are lined with hair and filled with a fluid. When you move your head, the fluid moves and the hair sends messages to the brain telling the body how to stay balanced. If you spin around in circles, the fluid and hair send messages. When you stop, they keep sending messages because they are still moving from having been shaken up. This is what makes you dizzy!

 **FUN FACT!** Earwax is made by the skin on the inside of the passageway in your outer ear. It helps keep your ears clean by catching dirt and dust. As you move your jaw when talking or eating, the wax is pushed outward.

 ## Special receptors in your skin allow you to feel things.

Underneath your skin are nerves that allow you to feel sensations. There are four types of special nerve receptors, for pain, pressure, cold, and heat. Receptors are located around the hairs in the skin. Because the receptors are close together, many sensations use a combination of receptors. For instance, if someone punches you in the arm, you feel pressure and pain.

Receptors can adapt, too. If you get into a steaming hot tub, you may only be able to put your feet in at first. Then, after your whole body has been in for a while, you get used to it. This is because the receptors have adapted to the temperature.

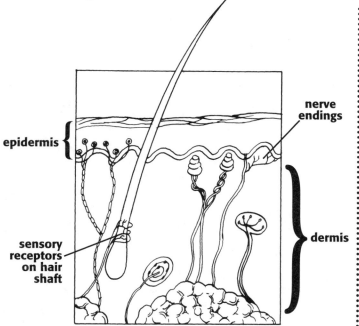

epidermis

nerve endings

sensory receptors on hair shaft

dermis

TOUCH AND EMOTIONS

You may not realize it, but touch is important to your emotional well-being. Babies that are not held and touched do not grow as well as they should—in fact, they can even die. Studies have shown that babies in the hospital who were not held regularly by nurses would gain less weight and cry more often than babies who were regularly held.

HANDS ON!

FIND YOUR PRESSURE RECEPTORS: With a friend, take a paper clip and unfold it. Shape it to look like a V. Blindfold your friend, and use the two points of the V to tap his or her fingertips. Does your friend feel both points of the V? Now tap your friend's arm. Again, see if your friend feels both points of the V. The arm may be less sensitive because our sense of touch is more precise on the fingertips.

70 Your **vocal cords** cause vibrations that make up your voice.

You can do many things with your voice—whisper, sing, hum, scream, and, of course, talk. When your brain sends a signal to make a noise or speak, your vocal cords respond. The vocal cords are bands of tissue at the bottom of the larynx, or voice box. When air comes out of the lungs, the vocal cords squeeze together to make a noise. The air vibrating on the vocal cords makes the sound. When you breathe and you aren't talking, your vocal cords are just relaxed.

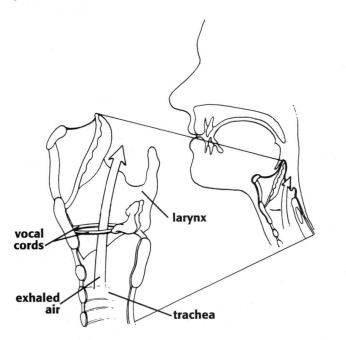

vocal cords

larynx

exhaled air

trachea

YOUR CHANGING VOICE

Children have short vocal cords, which cause higher sounds. As children grow, their vocal cords get longer, thicker, and farther apart. Then their voices become deeper. This change is more noticeable in boys because their voices generally get deeper than girls' voices. Often, there is a stage where a boy's voice will switch back and forth between low and high.

THE ORIGIN OF LANGUAGE

Linguists, people who study languages, believe that humans have always had some way to communicate with one another. Communication may have started with signs, pictures, and even simple sounds. Today, more than 6,000 different languages exist, and linguists believe they all grew out of one or just a few languages.

71 Your teeth are made of the hardest material in your body: enamel.

Next time you eat something, pay attention to how helpful your teeth are! Teeth help you chew food, and they add to your smile! Adults have 32 permanent teeth.

There are four types of teeth, each with a different job. Look in the mirror to see the types of teeth you have. In the very front of your mouth are eight **incisors,** which are pretty sharp teeth. These are used for biting. The four **canines** are the pointy teeth, two on top and two on bottom. They are used for tearing tough foods. There are eight **bicuspids** for crushing food. In the very back are 12 molars, used for grinding and mashing your food.

Although teeth are very hard, they are not hard all the way through. Like trees, your teeth have roots. Inside each tooth's crown, the outside part, a root holds the tooth into the gums. The crown is made of enamel.

BABY TEETH

It may seem like baby teeth, also called primary teeth, are not important because eventually they just fall out. But while they grow in, they help your jaw to develop properly. They also guide the permanent teeth in properly.

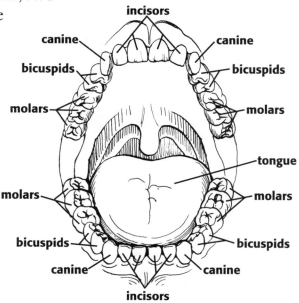

TAKING CARE OF YOUR TEETH

Brushing your teeth is something you should do after every meal. Flossing (using dental floss) can get the food between the teeth where a brush may not be able to reach. Brushing and flossing prevents tooth decay by killing germs that can feed on extra food in your mouth. Tooth decay eats through the enamel of your teeth. If the decay is deep enough, it will expose the nerves and can be very painful.

Enamel is the hardest thing in your body, even harder than your bones. Underneath the enamel is a layer of **cementum,** which covers the root and goes down into your gums. **Dentin,** which is like bone, forms most of the tooth and is beneath the cementum. The very inside of a tooth is the pulp, which is soft and contains blood vessels and nerves. The nerves allow you to feel hot, cold, and pain.

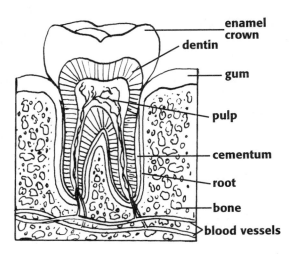

WISDOM TEETH

Wisdom teeth are way in the back of the mouth. They are called "wisdom teeth" because people get them in their late teens or early 20s, when they supposedly have gained more wisdom.

   Some people don't have wisdom teeth. Others have them removed because they cause pain. Wisdom teeth don't serve a purpose now, but scientists believe that our ancestors may have used them to help chew tough meat. Fossils of skeletons show that our ancestors had larger jawbones that stuck out quite a bit more than the jawbones of modern-day humans. The larger jawbones probably provided more room for wisdom teeth.

The digestive system breaks down food to fuel your body.

The digestive system starts in the mouth, where food is chewed. It ends at the anus, where food exits your body. In between, the stomach acts as a storage place that can hold up to 2 quarts of food. The stomach holds partially digested food for 3–5 hours. It slowly releases food to the rest of the digestive tract.

FUN FACT!

Your salivary glands produce about a quart and a half (6 cups) of fluid each day.

Food starts to break down the minute you bite into it. **Saliva,** a watery substance in your mouth, does more than wet your food so it will slide down your throat. As your teeth chomp away, a chemical in saliva breaks down the food.

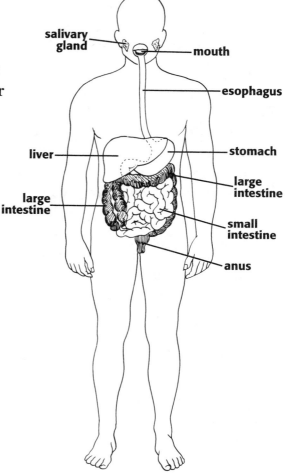

HANDS ON!

NOTICE STARCH AND SUGAR:

All you need are some plain crackers or a piece of bread. Take a small bite, and let it sit on your tongue for a moment. Does the taste change? It should get a little sweeter. That's because the saliva in your mouth has started to convert the starch in the crackers or bread to sugars.

SALIVA

Saliva is a fluid that comes from three pairs of salivary glands. Some are in your lips, cheeks, and tongue, but the larger salivary glands are just below your ears (on the inside, of course). Saliva helps your mouth stay healthy. It kills bacteria and helps keep your mouth clean by washing extra food away.

 # Food travels from your mouth to the pharynx and esophagus.

Once your food is chewed, it becomes a moist ball called a **bolus.** The bolus goes from your mouth to the pharynx in the back of the throat, where it is swallowed.

The pharynx leads to two tubes. One is the trachea, the windpipe to the lungs. The other tube is the esophagus, which goes down to the stomach. You may wonder how the food gets into the correct tube. The epiglottis is a valve that prevents food from getting into the larynx. It acts like a flap that closes down over your windpipe each time you swallow. After you swallow, the flap moves up to allow the air into the windpipe to get to your lungs. If food gets into the windpipe by mistake, you will choke. Your body has a reflex reaction to cough it out.

In the pharynx, automatic reflexes cause you to swallow. The food then moves into the esophagus, a muscular tube that pushes your food down to your stomach in seconds.

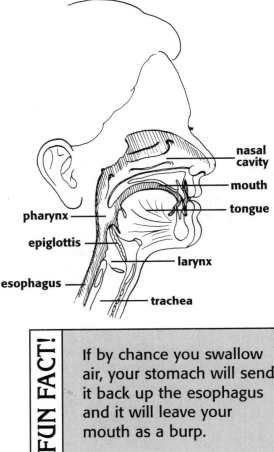

pharynx
epiglottis
esophagus
nasal cavity
mouth
tongue
larynx
trachea

FUN FACT! If by chance you swallow air, your stomach will send it back up the esophagus and it will leave your mouth as a burp.

THE HEIMLICH MANEUVER

If a piece of food gets stuck in someone's windpipe, it needs to come out. Usually a person's reflex reaction will cough the food out. But sometimes the piece of food will become so jammed in that a cough doesn't work. If this is the case, the person could die of suffocation because the food will block oxygen from getting into the lungs.

The Heimlich maneuver is a way to get the food out. A helper hugs the choking person from behind and puts a fist between the belly button and the ribs. Then he grabs his fist with the other hand and pulls, giving the person a quick jerk in the gut. This jerk may need to be repeated, but it should pop the food out into the air. If the person is lying down, the same technique can be used by pushing down on the same spot.

THE ADAM'S APPLE

Both men and women have an Adam's apple, a cartilage bump in the front of the neck. Men's Adam's apples are larger and more noticeable because male hormones cause the cartilage of the Adam's apple to grow more than in a woman.

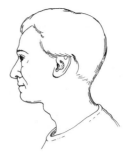

74 The juices in your stomach break down food and turn it into a pulp.

You may think that chewing your food makes it mushy enough to travel easily through your body. But in order for food to be absorbed into the bloodstream, it must be a liquid that can pass through the walls of the **small intestine** to the blood vessels. The stomach takes the chewed

FUN FACT!

The lining of your stomach has 35 million glands that release 2–3 quarts of stomach juices a day.

food (the bolus) and churns it. As the food moves around, it is mixed in with the stomach's **gastric juices**. Some of these juices are strong enough to dissolve metal! The food turns into a pulp called **chyme** and moves slowly to the small intestine.

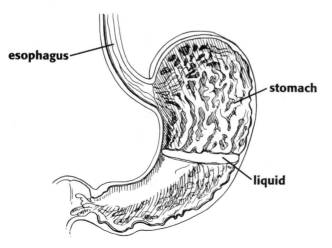

esophagus

stomach

liquid

VOMITING

Vomiting is an involuntary reaction to get rid of food you cannot digest. The reason for vomiting could be that you ate something bad, you are sick, or you ate too much. When you vomit, your diaphragm and stomach muscles contract to make strong jerks that push everything up through your esophagus.

STOMACH ACID

Could the stomach digest itself? This is a common question, since some acids in the stomach are strong enough to dissolve metal. But the stomach is coated with a thick layer of slimy mucus that prevents the acids from eating away at the lining. Also, once the acids are mixed in with the food, they are not as strong.

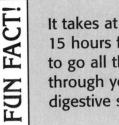

FUN FACT! It takes at least 15 hours for food to go all the way through your digestive system.

75 The small intestine, 18–23 feet long, absorbs nutrients into your body.

Both the small and the **large intestines** loop back and forth in your body. If they didn't, they would never fit! As partially digested food enters the small intestine through a tube at the end of the stomach, all sorts of things start to happen. The **gallbladder** releases **bile** into the small intestine to help break down fats. The pancreas and small intestine release **enzymes** to help break down other foods. It is here that the food is really turned to liquid so that it can pass through the intestinal walls and be absorbed into the bloodstream.

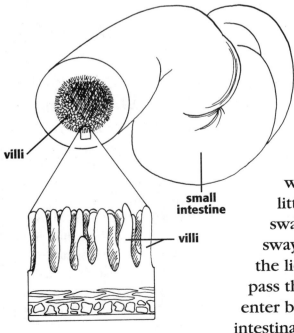

villi

small intestine

villi

How is it absorbed? Along the walls of the small intestine are tiny little **villi**. Villi are like little hairs that sway back and forth. While they are swaying, they grab on to the nutrients in the liquid, digested food. The nutrients pass through openings in the villi and enter blood on the other side of the intestinal wall.

76 The large intestine collects what is left over.

By the time the food makes it to the end of the small intestine and into the large intestine, digestion is nearly finished. Fiber (such as the fiber in vegetables), fats, water, and waste make it into the large intestine, from which they eventually exit the body through the anus.

About 95 percent of what you eat is used to fuel your body. Only about 5 percent of what you eat makes it to the large intestine. At the end of the small intestine, almost all the food you have eaten has been broken down and absorbed into the bloodstream to be used by your body. Leftovers from the small intestine enter the large intestine as a liquid. This liquid, called intestinal chyme, travels through the large intestine, and most of it is absorbed through the walls and recycled through the bloodstream. What is left can include proteins, dead bacteria, undigested food roughage (fiber), and water. All of these mix with mucus to form feces. Feces are eliminated from the body when you go to the bathroom.

77 The liver does over 500 jobs for your body.

The liver is the biggest organ inside your body, and it has many functions. Here are some of the things your liver does for you:

- Makes bile. This function is linked to digestion. Bile is a combination of water and salt that is released into the small intestine to break down fats.
- Processes nutrients. Once the blood has absorbed nutrients from the small intestine, it goes up to the liver. Basically, the liver is a dropoff point for all

CIRRHOSIS OF THE LIVER

Cirrhosis of the liver occurs when liver cells are damaged and replaced by scar tissue. Alcohol abuse and poor eating habits can cause liver damage. Hepatitis can also cause cirrhosis. Doctors do not fully understand this disease.

the nutrients. The liver takes the nutrients and separates them from one another. Some are saved for later use, and some are sent to places where they are needed.

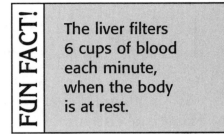

The liver filters 6 cups of blood each minute, when the body is at rest.

■ Filters. The liver, while it is sorting through all of the nutrients, recognizes things that aren't good for you, like poisons or waste products from the blood.

■ Controls **blood sugar** levels. The liver can turn sugars and fats into protein and protein into sugar. This is how it controls levels of sugar in your blood. Right after a meal when your blood sugar level is high, the liver takes sugar (glucose), changes it into a protein (glucogen), and stores it. When the blood sugar level goes back down, it takes the glucogen and turns it back into glucose.

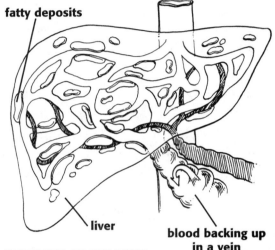

fatty deposits

liver

blood backing up in a vein

CIRRHOSIS OF THE LIVER

THE GALLBLADDER

The gallbladder is a storage place for the bile produced by the liver. By concentrating the bile, the gallbladder can hold all the bile the liver produces in half a day. The gallbladder releases some of the bile into the small intestine as the food gets there.

 **78** | # The pancreas helps you digest food and keep a regular blood sugar level.

The pancreas produces juices so strong that it could digest itself! But this would never happen unless the pancreatic juices were to back up instead of flow into the small intestine. The pancreas is part of two systems—the digestive system and the endocrine system.

For digesting food, the pancreas produces enzymes and releases them into the small intestine. Enzymes are chemicals that help break down food. The enzymes help digest proteins, carbohydrates, and fats. The pancreas also secretes an alkaline (sodium, or salt) solution to neutralize the stomach acids and keep them from eating away at the intestinal lining.

As part of the endocrine system, like other organs in this system, the pancreas produces hormones. One important hormone is insulin, which works with other organs to keep the blood sugar level in the body balanced.

DIABETES

The **islets of Langerhans** are small structures in the pancreas that produce the hormone insulin. The blood sugar level in your body is like a seesaw. When you eat, your blood sugar rises, and the seesaw tips. Insulin travels around the body and helps even out the level of sugar in your blood. The seesaw is balanced again.

pancreas

islet of
Langerhans

close-up of cells

If a person's pancreas doesn't produce enough insulin, the blood sugar level is out of balance, resulting in a disease called diabetes. Some diabetics have such a severe case that they must carry insulin and inject it when their blood sugar level gets too high.

79 Your brain takes a "reading" to tell you if you're thirsty or hungry.

How do you know you're thirsty? Your mouth might feel dry, or it might be hot outside and you've been running around sweating. Why you feel thirsty is pretty simple. You lose water, by sweating, going to the bathroom, even exhaling. When water leaves your body, you have to put water back in.

FUN FACT!

For good health you should drink eight glasses of water a day. If you play a sport or are outside in the heat, drink even more water.

Your body knows it's thirsty because it can "read" the water level in the blood. When the water level gets a little low, your brain takes a "reading" and can tell that more water needs to be added.

This is how hunger works, too. Blood is always running through your brain. Part of your brain keeps track of the nutrients in your blood. When the nutrients are low, you get a message: "I'm hungry! It's time to eat!" While your brain can tell when you need food most of the time, sometimes you won't get the message—such as when you're very sick or really busy.

A GROWLING STOMACH

When you hear noises from your stomach, chances are they are really coming from the small intestine. These "growls" are commonly believed to be cries of hunger from the stomach. They don't necessarily mean you are hungry, though. The sounds come from gas moving around in your digestive system.

SUMMER HUNGER

Why aren't you as hungry in the summer? Temperature has something to do with it. When it is hot and your body doesn't need as much fuel to keep you warm, you don't require as much food. If it's cold out and your cells work harder, you will be hungrier than usual.

 You fuel your body by eating from the six food groups.

We know that a body needs a balanced diet to make sure it has enough fuel to function well and to prevent certain illnesses. Balanced nutrition means eating from the different food groups: bread, grains, and pasta; fruits; vegetables; meats, fish, eggs, and beans; milk and dairy products; and a small amount of fats, oils,

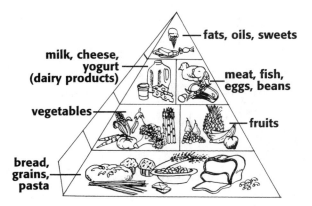

and sweets. Part of the reason you eat three meals a day is to meet your energy needs. You may not realize it, but you burn fuel even when you're sitting or sleeping!

NUTRIENT GROUPS

Foods are made of three main groups of nutrients that your body needs:

- Carbohydrates are fuel foods. Bread, cereal, pasta, and potatoes are foods high in carbohydrates.
- Proteins help your body repair and grow. Meats, eggs, cheese, and beans are high in protein.
- Fats are for energy. Cream, butter, and oil provide fat in the diet.

HANDS ON!

FIND STARCHES AND FATS: Find a couple of your favorite foods, and test them to see what nutrients they have. All you need is a piece of brown paper and some iodine. Checking for fat is easy. Just take a piece of food and press it to the brown paper. You will be able to see any fat that is left behind on the paper. You may want to hold it up to the light to get a better look. To check for starch (which is what carbohydrates have in them), put a drop of iodine on the food. If starch is in the food, the iodine will turn blue or black.

You can check for proteins, but you'll need to buy some chemicals or a chemistry set to do it. You'll need to mix a drop of potassium or sodium hydroxide with a few drops of water and copper sulfate solution. Then, put a drop of this mixture on the food. It will turn pink or blue if there is protein in the food.

SAFETY ALERT

Ask an adult to supervise this activity. Ask him or her for the iodine, potassium, and sodium hydroxide. Many households will already have iodine. You may need to buy the other two at a science-supply store.

81 Food contains vitamins and minerals required for good health.

Hundreds of years ago, when sailors relied on dried food for their long trips across the oceans, many of them developed a disease called scurvy. The foods they were eating did not have vitamin C, so their bones weakened, their teeth became loose, and many sailors even died. These are some of the results of scurvy.

A varied diet from the six food groups will provide all the vitamins and minerals your body needs to avoid such things. Vitamins and minerals are important substances in the foods you eat. When food goes through your digestive system, it is broken down until these nutrients are separated out and absorbed into your bloodstream.

HANDS ON!

KEEP TRACK OF YOUR VITAMINS AND MINERALS FOR A DAY: Not including vitamin pills, keep track of what you eat for a day. Write down the percentage of the Daily Values of vitamins and minerals you eat by consulting the nutrition label on food packages. If you eat out or at school, you will have to ask the person serving the food to show you the labels. Also, there are books you can use to find the nutritional values of certain foods. Do you eat the right amount of vitamins and minerals?

CALORIES AND ENERGY

A calorie is a way to measure the energy in food. If you read the label on any package of food, you'll find a listing for calories. Calories are a way to show how much energy food has to offer. If a serving of fruit has 100 calories and a serving of bread has 120 calories, the bread will have more energy. People who are watching their weight often count their calories. But the types of food and the fat intake you have also affect how your body burns energy. A more active person may burn more calories than a less active person.

82 Exercise is good for your muscles and your heart.

Exercise is important for everyone. You may get exercise when you walk to school, play a sport, or just run around with your friends. Once you're an adult, you may have to make sure that exercise is part of your lifestyle. For instance, many people whose work requires them to sit at a desk all day belong to a health club, a sports team, or a gym in order to get exercise.

Exercise helps your muscles by making them strong. It helps your bones because bones are growing all the time. When stress is put on your bones, they grow better. Running, for instance, puts stress on the bones in your legs. If you stop exercising completely, your muscles will start to get small and weak, and so will your bones! This is why people who are bedridden are in danger of having bones that may break easily.

Cardiovascular exercise works your heart. Exercising the heart by making it pump harder can help prevent heart disease in adults. A heart that is in good condition will beat more slowly. It will become more efficient. With each beat, it will carry more oxygen and blood to the body.

HANDS ON!

CHECK YOUR FITNESS: If you're in good shape, your heart rate recovers quickly after exercise. With a friend or two, see how your heart rates recover after a short sprint. First, take your pulse rates before starting. Time your pulse for 15 seconds, then multiply by four to get the pulse rate per minute. Then sprint for at least 100 yards. Take your pulse rates immediately upon stopping, then every 30 seconds afterward. How long does it take for your pulse to get back to its normal rate?

83 Your weight varies according to your body type.

If you flip through any magazine, chances are you'll see models—especially women—who are extremely thin. Although magazines and television tell the world that "thin is in," it's not really what everyone should look like. In fact, there are many different body types, and when doctors say what weight you "should" be, there

FREAKY FACT! 64 percent of Americans are considered overweight.

is a range. For instance, a healthy weight for an adult woman who is 4'8" tall is anywhere from 96 to 125 pounds. Two people may weigh the same, but one person may be smaller in size than the other. This could have to do with how the bodies are shaped and who exercises more, because muscle weighs more than fat.

In general, a person who is 20 pounds over the suggested weight range is considered overweight. Being overweight is not healthy for anyone. It puts extra stress on the heart.

FUN FACT! The average height for an adult woman is 5'3¾". The average height for an adult male is 5'9".

EATING DISORDERS

Anorexia nervosa and bulimia are **eating disorders** caused by an intense fear of being fat. Anorexia can cause a person to starve her- or himself—even to death. Bulimia is a disorder in which the person binges (eats a lot) and then vomits. Bulimics often become depressed and can also die from their eating disorder.

People most at risk of developing these disorders are females between the ages of 12 and 14. Doctors are not sure about the reasons behind eating disorders. But eating disorders may indicate that the person does not feel very good about her- or himself and does not have much control over her or his own life. The person may feel more attractive by being severely thin. Some people with eating disorders may look in the mirror and see a distorted, heavier image that "needs" to lose weight.

84 The kidneys and bladder of the urinary system rid your body of waste.

Kidneys are bean-shaped organs located in your lower back. The kidneys, even though they are part of the urinary system, have a great impact on the circulatory system. Here are the main functions of the kidneys:

- Filter blood and remove waste. The kidneys are like a filtering system for the blood. The blood is pumped into the kidneys from the heart. The blood goes into nephrons, tiny little filters that know how much water and nutrients should stay in the blood. These filters also separate what the body can use from waste. Any extra water and waste are turned into urine. The urine trickles down to the bladder. When your bladder is full, you get a message that it's time to go to the bathroom.

- Monitor blood pressure. If a person's blood pressure is too high, the kidneys will get rid of more fluid in order to decrease the amount of fluid in the body. Less fluid means less pressure.

- Monitor red blood cells. If the cells in the body are not getting enough oxygen, the kidneys react. They release a hormone that tells your body to make more red blood cells. With more red blood cells, more oxygen can be delivered.

FUN FACT! The kidneys receive about a quart of blood a minute.

DEHYDRATION

A person can go for days without food, but not without water. In fact, a person could not survive even a week without water. Usually, people do not get into such extreme situations. Sweating a lot from being sick or from too much exercise can cause **dehydration** if you don't drink water to make up for the loss. Drinking tea, coffee, or soda will not help hydrate a person because these are diuretics, which increase the amount of urine the kidneys produce.

FUN FACT! The amount of water in your body remains pretty much the same all the time. The amount you take in is the same amount you get rid of.

85 Your bladder muscles help you control when you go to the bathroom.

Urine is released when the bladder fills with about 2 cups of urine. Nerves in the bladder send a message to the brain. The brain tells the bladder muscles to contract, and the sphincter muscle lets the urine pass. Urine leaves the body through a tube called the urethra.

Controlling the urinary muscles is learned in most children around the age of 2½. Before this age, urinating is a reflex action, and the part of the brain that controls these muscles has not developed.

FUN FACT! Urine is about 95 percent water and about 5 percent waste.

WATER AND THE HUMAN BODY

Many ingredients make up the human body, but the main one is water. Water makes up 70 percent of your body! But the water is a part of many things. For instance, there is water in your blood, which is circulating all around inside of you. Your bones are one-third water. Your liver produces bile, which is mostly water, too.

Water comes into your body when you drink it. It is distributed by the digestive system. Water leaves when you go to the bathroom. It also escapes through your skin when you sweat and through your breath when you exhale.

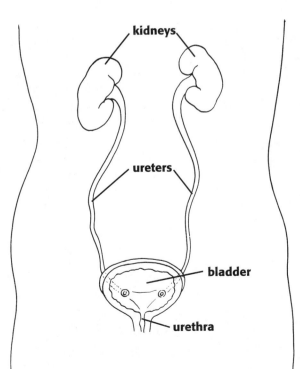

kidneys

ureters

bladder

urethra

TESTING 1, 2, 3

A lot can be learned from a person's urine. In ancient times, urine was one of the main tests for patients. Its color, smell, and taste would help early Greek doctors tell what could be wrong with a patient. Today doctors look at urine samples under a microscope to look for bacteria. Bacteria can show whether a kidney is performing well or not. A doctor can also see if there is too much sugar in the urine, which could be a sign of diabetes.

86 You have two kinds of glands in your body: exocrine and endocrine.

Glands, located throughout your body, fall into two groups. **Exocrine glands** get their name from the Greek word *exo,* which means external or outside. Exocrine glands secrete chemicals into a duct (pathway) that carries them to the outside of the body or to a body cavity. The salivary glands are exocrine glands that release saliva into your mouth. Sebaceous glands are located all over your skin and release oil to keep your skin soft. Sweat glands let sweat out through the pores of your skin.

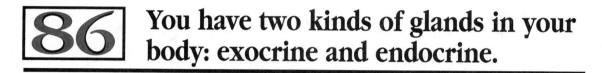

salivary glands

Endocrine glands get their name from the Greek word *endo,* which means internal or inside. Endocrine glands secrete hormones right into the bloodstream. The bloodstream transports the hormones to where they are needed in the body.

87 The endocrine system is made up of endocrine glands that release hormones.

The endocrine system is not as familiar as the other body systems, but it is very important. Endocrine glands produce hormones, which regulate such processes as growth, digestion, and body temperature control. Several main glands, and hundreds of smaller glands, make up the endocrine system.

Glands are like factories that produce hormones. Hormones are chemical messengers that travel around your body and send signals. They are transported around the body through the bloodstream. Your body makes over 100 different kinds of hormones, each of which does a different job. Messages from hormones know which parts of the body to go to.

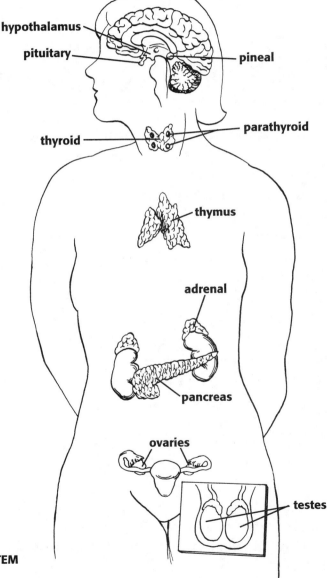

ENDOCRINE SYSTEM

88 | The **pituitary gland** is the "leader" of the glands.

The pituitary gland is about the size of a pea. It is located in the brain, in the center of the skull, right behind your nose. It is sometimes called the master gland because it releases hormones that tell the other glands what to do.

By the age of 5, you had probably already grown to half your adult height. That is pretty amazing, considering you still have about 15 growing years left! Growth slows after the 5-year mark, but right before your teens, you have another growth spurt. How does your body know when to grow? One hormone that comes from the pituitary gland is **human growth hormone (HGH)**. This hormone signals your bones to grow. Even though you do not keep getting taller after about the age of 20, HGH is produced throughout your life. When you stop growing, HGH helps the body process food. Other growth hormones are responsible for the growth of your muscles and other organs.

89 | The **thymus gland** and the **pineal gland** are known as mystery glands.

Even though the thymus produces immune system cells in babies, doctors are not completely sure of the gland's function. It is studied as part of the endocrine system, but so far no one has discovered what, if any, hormone the thymus produces. Scientists are not sure what the pineal gland does

THE THYMUS GLAND

A newborn baby has an undeveloped immune system. The immune system develops shortly after birth. Meanwhile, the thymus gland produces T cells. The thymus gland is large in a newborn, but by age 10, it has shrunk down to about the size of a thumb. Scientists think the thymus may provide young children with extra protection against diseases.

either. This tiny gland, located in the head, may have something to do with the control of sleeping habits. It also may be related to **reproduction**.

 # Your body has an internal clock.

Your internal clock, also called a circadian rhythm, is thought to be controlled by the pineal gland. Imagine staying in an isolated room with no sunlight, no alarm clock, and no television. If you stayed there for a long time without being able to communicate with other people, how would you know when to sleep and wake up? Your internal clock would tell you.

Researchers conducted a study in which they isolated a number of people for about a month. They stayed in underground rooms and could not see the sun rise and set. They had no alarm clocks or television, no information from the outside world. The researchers told the subjects to sleep when they wanted and to try and avoid taking naps. The results showed that each subject had a pretty regular sleep schedule. The one conclusion

FUN FACT!

The word *circadian* is Latin and means "about a day."

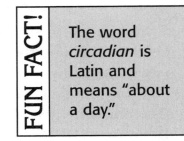

HANDS ON!

KEEP A SLEEP LOG: For a week or so, write down how many hours of sleep you get each night. Also keep notes on how you feel on certain days. Do you sleep the same amount of hours every night? Do you sleep more on the weekends? How do you feel on the days that you get more sleep?

JET LAG

You probably have a daily routine. You get up at a certain time to go to school. You eat your meals at about the same times each day. You may even go to bed around the same time each night. Your internal clock gets used to a regular schedule like this. When you take a trip on an airplane and cross into another time zone, you may experience **jet lag** from having your regular schedule disturbed.

from the study was that the awake time and sleep time of the subjects averaged 25-hour-long days. Even though our days are 24 hours long, our bodies may be living on a 25-hour day!

91 The **adrenal glands** help you respond to fear and stress.

If you have ever been frightened by someone jumping out to surprise you, you may have experienced an adrenaline rush. When you experience fear, anger, nervousness, or stress, the adrenal glands respond by releasing the hormone adrenaline into your blood. Your heart will beat faster and your blood pressure will rise. You'll feel like you have more energy. Why? Your body is getting ready to make a run for it, or to fight whatever it is that has just scared you. This is often referred to as fight-or-flight response. Of course, once you realize that your friend has just played a joke by jumping out at you, you calm down. But in other situations in which real danger exists, your adrenal glands will help you respond.

SUPER-STEROIDS

Certain hormones produced by the adrenal glands are called steroids. Steroids help control the level of salt, sugar, water, potassium, and other substances in the blood. But steroids are also used to treat people who have asthma attacks, or people who are stung by and are allergic to bees.

92 The **thyroid gland** helps you have energy.

You eat food for fuel, and the fuel is turned into energy. But how does your body know how quickly to burn the fuel? The thyroid gland controls the speed at which fuel should be burned. Food goes through your digestive system and gets broken down into a liquid. Eventually nutrients from the food are absorbed by the blood and carried to all the cells in your body.

The cells take in the nutrients (with oxygen) and make energy out of them. The thyroid gland produces hormones that tell your body how quickly or how slowly to do all of this. If your body makes energy too quickly, you will lose weight. If your body makes energy too slowly, you will feel lazy and probably gain weight.

GLANDS FOR SURVIVAL

On the sides of the thyroid gland are tiny glands called parathyroid glands. We could not survive without the hormone that these glands produce. The hormone, called **parathormone,** helps regulate calcium levels in the blood. Calcium is important for building healthy bones and teeth.

ENERGY FROM CELLS

As cells burn fuel, they create energy called **ATP**. Active people have cells that burn fuel fast and make more ATP. If you could get the ATP from your cells after they have burned 3,500 calories of food, the energy would be enough to turn on 1,500 100-watt light bulbs for 1 minute.

93 Sex hormones and sex organs are the basic differences between men and women.

Men and women have many things in common. In school, boys and girls study the same topics. They participate in many of the same sports. In the working world, men and women compete for the same jobs. But men and women play very different roles in human reproduction.

Around the age of 12, the bodies of boys and girls go through a growth spurt and start to change. This time is known as puberty. Boys get deeper voices and hair on their chests, armpits, and genital areas. Girls' hips become rounded and their breasts start to develop. These changes happen because the pituitary gland signals the reproductive system to release sex hormones. In boys, the hormone is called **testosterone** and is produced in the **testes**. The testes are small sacs behind the **penis** that start to produce sperm at this time. In girls, the pituitary gland commands the ovaries to produce the hormones **estrogen** and **progesterone**. The ovaries contain eggs. The release of these sex hormones is one way the body gets ready for reproduction. The hormones cause the physical changes and help boys and girls develop into men and women.

TESTOSTERONE AND ESTROGEN

Even though testosterone is considered a "male hormone," testosterone is also produced in females. Estrogen, considered a "female hormone," is also produced in males. The difference is in the amount of hormones for each sex: women have much less testosterone than men, and men have much less estrogen than women.

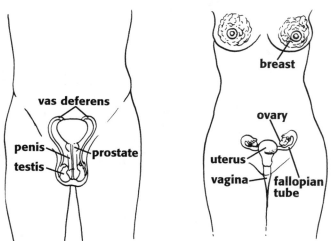

MENSTRUATION

At the time of puberty, which is when the sex hormones estrogen and testosterone are produced, a young woman starts **menstruating**. Each month, an egg is released from the ovaries. This is called ovulation. As the egg travels down the **uterus,** the walls of the uterus become thicker because of a hormone that is released during ovulation. If the egg does not become fertilized, the hormone release stops and the thick lining of the uterus starts to shed. It passes through the **vagina** in a flow of blood during the menstrual period. Each period lasts for about 5 days. The cycle repeats every 28–30 days, unless a woman becomes pregnant.

 **The main parts of the male reproductive system are the penis and testes.**

The function of the male reproductive system is to produce and deliver sperm cells to the female. A man's reproductive organs are mostly on the outside of his body. The sperm, which are needed to fertilize a female's **egg**, are produced in the testes. The testes are two small sacs that hang behind the penis. The testes are outside the body to provide a cooler place for the sperm, because sperm cells would not survive in a warmer environment. An adult male produces 100 million sperm a day.

The penis is long and tube-shaped. The shape allows sperm to be released close to the eggs during sexual intercourse. Millions of sperm are released in about a spoonful of semen, a liquid that comes out of the man's penis. The sperm swim upward toward the egg to fertilize it. Many sperm are killed by white blood cells, and others die. Only the strongest and fastest sperm will make it to an available egg.

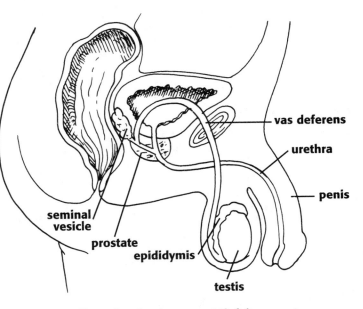

vas deferens

urethra

penis

seminal vesicle

prostate

epididymis

testis

95 The main parts of the female reproductive system are the ovaries, uterus, **fallopian tubes**, and vagina.

The female reproductive system provides eggs for **fertilization** by the male's sperm. While a man's sex organs are mostly on the outside of the body, a woman's are on the inside. In addition to providing eggs, a woman provides a place for a fertilized egg to grow into a baby and live there for 9 months!

Each month, an egg will make a journey starting in the ovaries. When an egg is mature, it travels from the **ovary** down a long tube called a fallopian tube to the uterus. If the egg is fertilized and a baby grows, it will grow in the uterus. The uterus ends at the cervix, which opens up to a channel called the vagina. The vagina is where a man places his penis during sexual intercourse. It is also where the baby comes out during childbirth.

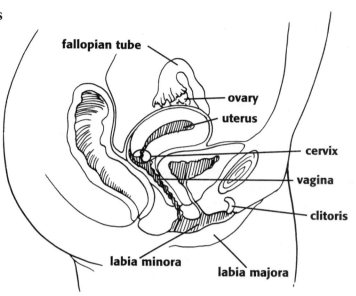

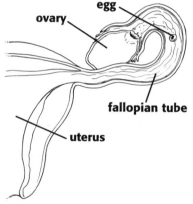

A woman's breasts are also considered part of the female reproductive system. After a baby is born, the mother's breasts start to produce milk for the baby to drink.

FUN FACT! A woman's egg is smaller than a pinhead.

96 The reproductive systems of men and women allow humans to make babies.

Men and women have sexual intercourse when a man puts his penis into a woman's vagina. The penis releases millions of sperm inside the woman, but only

FUN FACT! The story that storks deliver babies comes from a belief Northern Europeans had in ancient times. They thought storks delivered the souls of unborn children to people's homes.

one sperm can fertilize an egg. Once the sperm and egg meet, a one-celled **zygote** forms.

Fertilization depends on timing. The egg in the woman's body should be in the fallopian tube at the exact time a sperm finds the egg. An egg travels into one of these tubes once a month and lives for about 1 day. Sperm live from 2 to 4 days.

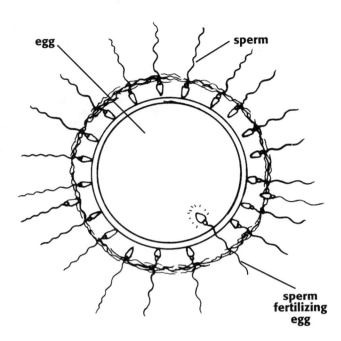

egg

sperm

sperm fertilizing egg

97 | Chromosomes from each parent determine what you will look like.

When a father's sperm meets a mother's egg, each brings 23 chromosomes. The first cell formed by the sperm and the egg has 46 chromosomes (23 pairs). Before the cell splits, it copies the 46 chromosomes, so that when the second cell is made, it also has the 46 chromosomes. Every cell in the human body (except sperm and egg cells) contains 46 chromosomes.

Chromosomes contain genes, which determine individual traits—such as the color of one's hair and eyes and one's height. This is why you look a little like both of your parents. Two parents' chromosomes can be combined in so many ways that there is almost no chance of the same combination happening twice, except in the case of identical twins. This explains why brothers and sisters look similar, but not exactly alike.

98 You get your looks from dominant genes.

You inherit genes from your parents, but your traits really depend on the combinations the genes come in. Genes come in pairs, and there are two kinds: dominant genes and recessive genes. Recessive genes are like hidden genes. Dominant genes are the ones that always show. So if you inherit a pair that has one of each, the dominant trait is what you will get. The only way you can inherit recessive traits is if you get a pair of genes that are both recessive. One of your parents may have blue eyes, while the other parent has

BRIEF Bio

GREGOR MENDEL (1822–1884): Mendel was an Austrian monk who first understood and explained the principles of dominant and recessive genes. He explored how traits are handed down from one generation to the next by studying pea plants. He would cross-pollinate different kinds of peas to see which traits were passed down. By doing this, he could learn about dominant and recessive genes. Because of Mendel's research, we now know how some traits are passed on from the previous generations in plants and animals.

brown. Which one did you inherit? The gene for brown eyes is dominant. The gene for blue eyes is recessive. You probably inherited the brown eyes: a dominant brown-eye gene and a recessive blue-eye gene.

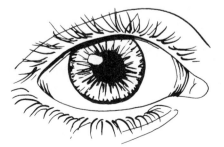

Say two parents both have brown eyes, and each carries one recessive blue-eye gene and one dominant brown-eye gene. Here are the combinations of eye colors their children could have:

- If the child inherits two dominant brown-eye genes: brown.
- If the child inherits one dominant brown-eye gene and one recessive blue-eye gene: brown.
- If the child inherits two recessive blue-eye genes: blue.

99 It takes 9 months for a fertilized egg to grow into a baby.

Remember that you started out as just one cell. This cell formed when one of your father's sperm cells penetrated one of your mother's eggs. The egg was fertilized and became a single cell called a zygote.

A zygote splits 30 hours after it is formed. As the embryo develops from the zygote, it is

FUN FACT!
After a baby is born, the umbilical cord is cut off, leaving the belly button in its place.

BEFORE YOU'RE BORN

Your very first 9 months were spent underwater. The fetus develops in a capsule of liquid that protects it. While you were in your mother's uterus the fluid came into your lungs, but you didn't drown. How did

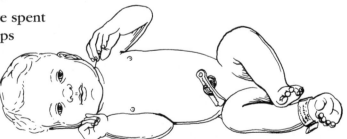

you get oxygen and food for all those months? The fetus gets oxygen and food from the **umbilical cord.** The umbilical cord goes from the stomach of the fetus to the **placenta,** in the wall of the mother's uterus.

The placenta is a spongy tissue that is like a gateway from the mother to the baby. The mother's blood never goes directly into the baby. Instead, the placenta passes nutrients and oxygen through the umbilical cord to the baby.

surrounded by a sac of liquid. This sac is called the amniotic sac. The liquid is like a padding of protection. The embryo will live in this sac for 9 months. It isn't until 4 weeks after that first cell splits that a heart is fully formed and starts to beat. After a whole month, the embryo is just starting to form fingers and toes. At 2 months, the embryo is only an inch long. At the ninth week, the embryo has all of its parts and is called a fetus.

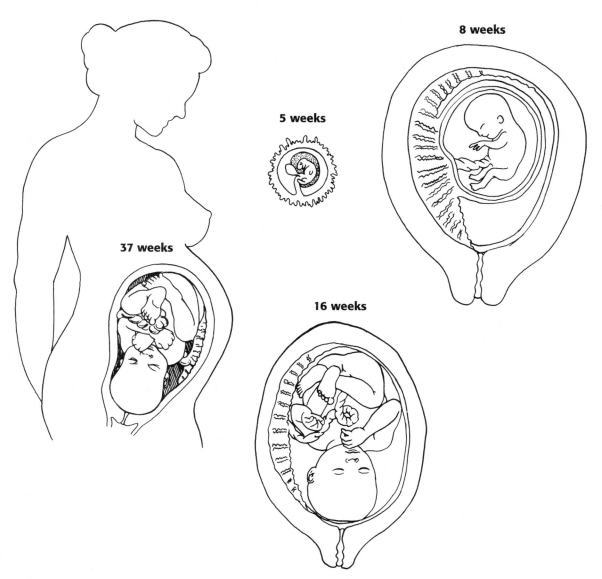

100 | Most people are full grown by the age of 20 and get shorter in old age.

From birth to old age, a lot happens to the body. Of course, your looks change, but other things change, too. A person grows for the first 20 years of life, but even after that there are changes.

- At one year old, most babies are just starting to learn how to walk.
- By age five, a child starts school, where reading and writing are learned.
- Puberty, a time of many body changes, starts for girls around the age of 11 or 12 and for boys around the age of 14.
- At 20, most people have done all of their growing, at least in height. Weight can still change depending on a person's lifestyle and heredity.
- Between the ages of 40 and 50, wrinkles may start to appear. By this time, some men and a few women have started to lose their hair, and some people will put on a little weight. As people get older, they could be more prone to illness.
- From age 60 on, the skin becomes more wrinkled and most people become shorter. Elderly people are about 2 to 3 inches shorter than they were at age 20. This happens because the discs between the bones in the back shrink.

101 | A person's average life span is 85 years.

Some people live to be over 100 years old, but few people live beyond age 85. Some scientists think that we are "programmed" to live a certain amount of time, the same way we are "programmed" to look a certain way. Our genes may carry the instructions for how long we are to live.

But even if your life span is determined by genes, environment has something to do with it, too. For instance, a person may have inherited genes that will allow him or her to live to the age of 90, but if that same person smokes and drinks heavily, his or her chances of living that long decrease.

After the age of about 80, the functions of the brain, muscles, joints, eyes, and ears decline. The best way to slow down the aging process is to maintain a healthy weight, exercise regularly, and avoid alcohol and smoking. Also, doctors recommend getting regular medical checkups to detect health problems early.

GLOSSARY

acid A strong substance that can eat away at solids.

adrenal gland A gland that produces the hormone adrenaline, which can cause the heart rate to escalate in fight-or-flight responses brought on by emotional distress.

AIDS Acquired immunodeficiency syndrome, a disease caused by HIV, a virus that is spread by sexual intercourse or infected blood.

albino A person who has no coloring in the skin or hair. Albinos have pink eyes.

allergy The body's reaction to harmless substances that are seen as invaders, such as plant pollen and dust.

alveoli Tiny air sacs in the lungs.

anemia A lack of red blood cells.

antibody A protein that protects the body by attacking microscopic invaders.

artery A blood vessel that carries blood away from the heart.

ATP A type of energy produced in the cells.

atrium One of two upper chambers of the heart.

axon An extension from the cell body of a neuron; carries outgoing signals.

bacteria One-celled organisms. Some kinds of bacteria can cause disease.

B cell A type of white blood cell that makes antibodies.

bicuspids Eight teeth, four on each side of the mouth (behind the canines and before the molars), used for crushing and chewing food.

bile A fluid produced by the liver to help with the digestion of fats.

blood A liquid that travels through blood vessels, carrying oxygen and carbon dioxide.

blood sugar The glucose (or carbohydrate) in the blood.

body system A group of organs and tissues that work together to perform tasks to keep the body running smoothly, such as the respiratory system.

bolus A ball of chewed food that has mixed with saliva.

brain The command center of the nervous system.

brain stem The part of the brain that connects with the spinal cord; controls most body functions.

bronchiole A smaller tube that branches off a bronchus.

bronchus A large tube branching off the trachea that passes air to each lung.

calcium A mineral that keeps bones and teeth healthy.

canine One of four pointed teeth near the front of the mouth.

capillary A tiny blood vessel that connects arteries and veins.

cardiac muscle A type of muscle that makes up the heart.

cartilage A tough material at the ends of bones that cushions them at joints.

cell The smallest unit of all living things.

cementum A thin, hard material that covers the outside of the root of a tooth.

cerebellum An area in the middle of the brain that controls movement and balance.

cerebral hemisphere One of two sides of the cerebrum, the part of the brain that controls thinking and learning.

chromosome A threadlike structure containing genes, which determines a person's traits.

chyme Partially digested food going from the stomach to the small intestine.

cone A special type of cell in the retina of the eye that enables one to see color and bright light.

contagious disease An infectious disease that is passed from person to person.

cornea The outer part of the eye covering the iris and pupil.

coronary artery One of two main arteries that supplies blood to the heart.

cytoplasm The material between the nucleus and cell membrane.

dehydration A lack of liquid (water) in the body.

dendrites The fine branches of nerve cells that carry incoming signals toward the cell body.

dentin Bonelike material that comprises most of a tooth.

dermis The layer of skin beneath the outer layer (epidermis).

diaphragm A sheet of muscle below the lungs that helps a person breathe. When the diaphragm contracts, it causes air to flow into the lungs.

eardrum The part of the ear that vibrates in response to sound.

eating disorder A disorder in which a person vomits, binges, or demonstrates some other behavior to try to lose a lot of weight; examples include anorexia nervosa and bulimia.

egg The female sex cell.

embryo The developing baby, from the third to the eighth week of pregnancy.

enamel A hard substance that covers the crown of a tooth.

endocrine gland A type of gland that secretes hormones into the bloodstream.

enzyme A protein that speeds up chemical reactions in food and helps break the food down quickly.

epidermis The top layer of skin.

esophagus The tube between the pharynx and the stomach.

estrogen A hormone produced by the ovaries that causes physical changes in girls during puberty; it also prepares a woman's body for pregnancy.

exocrine gland A type of gland that passes secretions to the surface of the body, such as sweat glands.

fallopian tube One of two tubes that extend from the ovaries to the uterus.

fat A necessary compound that fuels the body, found in energy-rich foods. While some fat is needed in a person's diet, too much fat can cause serious illnesses.

fertilization The joining of an egg and a sperm.

fetus The developing baby, from the ninth week until birth.

fiber A structure that resembles a thread; it can be a strand of nerve tissue, or a cell that is threadlike. Fiber is also found in foods such as fruits, vegetables, and grains.

gallbladder A small, baglike organ in which bile from the liver is stored.

gastric juice A mixture of acid and enzymes in the stomach to help with digestion.

gene A section of a chromosome that determines a trait; genes are inherited from both parents.

gland An organ that produces substances for the body to use.

hair follicle A tiny tunnel in the skin from which hair grows.

heart The organ that pumps blood through the body.

HIV Human immunodeficiency virus; the cause of AIDS.

hormone A chemical substance made in the body that triggers a specific response, such as growth.

human growth hormone (HGH) A hormone that controls growth in children and young adults; in adults, HGH helps the body process food.

hypothalamus A part of the brain that controls functions such as body temperature, heart rate, blood pressure, hormone release, and memory.

incisors The front eight teeth, four top and four bottom, that have cutting edges sharp enough to chop off bits of food.

infectious disease A disease caused by a bacterium or virus; can be caught from another person or from something in the environment.

intelligence The ability to deal with new situations, learn, and understand things, and to reason.

invertebrate An animal without a backbone.

iris The colored part of the eye; controls the size of the pupil, which controls the amount of light that reaches the retina.

islet of Langerhans A type of cell in the pancreas that secretes insulin.

jet lag Tiredness caused by traveling through different time zones.

joints A place of connection between bones that allows free movement.

keratin A protein that forms hair and fingernails.

large intestine A tube in the digestive system that does not carry out digestion but absorbs water and concentrates waste.

larynx The voice box at the opening of the trachea into the lungs.

lens The part of the eye that focuses light.

ligament Connective tissue that joins bone to bone.

liver An organ that deals with nutrients from digestion, stores energy (glucose and starch) and vitamins, and produces bile.

lung An organ that takes in oxygen for the body; the place where gas exchange occurs in the respiratory system.

lymph node A small gland packed with white blood cells.

mammal A vertebrate animal that nourishes its young with milk and has skin covered by hair. Humans are mammals.

melanin A pigment produced in skin cells that provides color and absorbs the sun's ultraviolet rays.

membrane A thin, skinlike layer with blood vessels and nerves running through it. Some types of membrane are found in bones; other membranes line the body's tubes and cavities.

meninges The three layers of tissue around the brain.

menstruation The monthly female reproductive cycle in which an egg is released.

microorganism A tiny organism, visible only with a microscope, that lives in the human body, and also in water, plants, etc. Examples include bacteria, yeasts, and fungi.

mineral A man-made substance, such as iron or calcium, that the body needs to stay healthy.

mitochondrion The part of a cell that generates energy.

mitosis The splitting apart of a cell to make two cells, each of which is exactly like the original cell.

molecule The smallest chemical unit of matter.

mucus A slimy liquid that protects the body by coating certain areas, such as the stomach lining and nasal passages.

nerve A bundle of axons that passes signals to the brain.

neuron A nerve cell.

neurotransmitter A chemical that sends a message at the synapse between two neurons.

noninfectious disease A disease that is inherited or comes from within the body. It cannot be caught from other people.

nucleus The part of the cell that controls how the cell operates.

nutrient A substance from food required for the body to function.

organ A group of tissues that act together to perform a specific function.

osteoblast A type of cell that builds bone.

osteoclast A type of cell that breaks down old bone.

osteocyte A mature bone cell.

ovary One of two organs in the female reproductive system that contain eggs.

pancreas An organ that makes digestive enzymes and hormones that control blood sugar levels.

parasite An organism that lives on another animal (including humans).

parathormone A hormone that controls calcium levels in the blood.

penis The male external sex organ.

phagocyte A type of white blood cell that destroys invading cells.

pharynx The passageway from the nose to the back of the throat to the esophagus.

pigment A substance in hair, skin, and the irises that gives color.

pineal gland A tiny gland located in the head that may control sleeping habits.

pituitary gland A gland in the brain that releases hormones that trigger the activity of other glands.

placenta Tissue that surrounds the fetus during pregnancy and links the blood supply between mother and baby through the umbilical cord.

plasma The fluid part of the blood.

progesterone A female sex hormone that prepares the female body for reproduction. Progesterone is also produced by men.

protein A compound in food that helps build muscle.

pulse A regular wave of pressure through the arteries, caused by the heartbeat.

pupil The hole in the iris that controls the amount of light that reaches the retina.

reflex An automatic response, such as jerking the hand away from something hot.

regenerate To make again.

REM (rapid eye movement) sleep A stage of sleep in which the eyes flicker back and forth under the eyelids and during which dreams occur.

reproduction When a male and female have sexual intercourse to create a new human being.

retina Layers of cells in the back of the eye that translate what you see into nerve impulses that travel to the brain, which interprets them as images.

rod a special type of cell in the retina of the eye that enables seeing in dim light.

saliva A watery fluid secreted into the mouth to help with digestion.

sinus A cavity (hole) in the skull.

skeletal muscle A type of muscle responsible for voluntary movements.

sleeping disorder A disorder that disturbs normal sleep habits.

small intestine A coiled tube where most of the digestion of food takes place. Nutrients from food enter the bloodstream through the walls of the small intestine.

smooth muscle A type of muscle responsible for involuntary movements, such as digestion.

sperm The male sex cell.

spinal cord A bundle of nerve cells that runs from the brain through the inside of the spine.

spine A column of 33 ringlike bones, also called vertebrae.

stomach A stretchy, muscular organ where food gets churned into chyme during digestion. Acids in the stomach help with digestion, and also help to kill any harmful bacteria in the food.

synapse The space between two neurons.

taste buds One of many tiny structures on the tongue that sends messages to the brain about the four basic tastes: sweet, sour, salty, bitter.

T cells Cells that kill invading germs and help control the immune system.

tendon A tough band of tissue that connects muscle and bone.

testes Two small structures that produce sperm.

testosterone A male sex hormone produced by the testes that prepares the male body for reproduction. Testosterone is also produced by women.

thalamus A region of the brain that processes information from the senses and relays it to the cerebrum.

thymus gland A gland that produces T cells of the immune system.

thyroid gland An endocrine gland that controls heat and energy production by improving the performance (metabolic activity) of cells.

tissue A group or layer of cells that acts as a unit.

trachea The windpipe to the lungs.

umbilical cord The cord between the mother's placenta and her developing baby.

uterus The female organ in which the embryo and fetus develop.

vagina The external female sex organ; a channel that leads from the uterus to the outside of the body.

vein A blood vessel that carries blood toward the heart.

vertebrate An animal with a backbone.

villi Microscopic, hairlike projections in the small intestine that help with the absorption of nutrients.

virus A particle that can live in body cells, reproduce itself, and cause an infectious disease.

vitamins Substances needed by the human body for nutrition that we get by eating natural foods. Vitamins help with chemical reactions in cells.

vocal cords Bands of tissue at the bottom of the larynx (voice box) that squeeze together to make sounds as air moves through them.

waste Leftover material that a body does not digest or will not use.

zygote The cell produced when an egg is fertilized by a sperm.

INDEX